BARRON'S

Regents Exams and Answers

Geometry

Revised Edition

T0165080

ANDRE CASTAGNA, PH.D.
Mathematics Teacher
Albany High School
Albany, New York

Kaplan North America, LLC d/b/a Barron's Educational Series
1515 West Cypress Creek Road
Fort Lauderdale, Florida 33309
www.barronseduc.com

ISBN: 978-1-5062-6634-3

10 9 8 7 6 5

Kaplan North America, LLC d/b/a Barron's Educational Series print books are available at special quantity discounts to use for sales promotions, employee premiums, or educational purposes. For more information or to purchase books, please call the Simon & Schuster special sales department at 866-506-1949.

Dedication

To my loving wife Loretta, who helped make this endeavor possible with her unwavering support; my geometry buddy, Eva; and my future geometry buddies, Rose and Henry, and my professional colleagues for their patience and understanding.

Contents

Regents Examinations, Answers, and Self-Analysis Charts   241

The Geometry Learning Standards   485

Preface

This book is designed to prepare you for the New York State Geometry Common Core Regents and is aligned to the Common Core Geometry Learning Standards. The key features you will find in this book are:

- **Test-Taking Tips** Advice and test-day strategies are provided to help you earn as many points as possible on the exam.

- **Brief Review of Geometry Concepts** A concise summary of the theorems, formulas, and math skills needed for success on the Regents exam is provided. Much of the information is presented in bulleted or table form to make it easy for you to use as a refresher and study aid. This summary is intended to supplement the more thorough presentation of high school geometry found in the Barron's *Lets Review: Geometry* book.

- **Glossary** A full glossary is found at the end of the Brief Review section. Be sure to look up any word that you are unfamiliar with as you work through the practice problems. Geometry is a vocabulary-intensive subject.

- **Actual Regents Exams and Step-by-Step Solutions with Explanations** Each of the problems in this book has a detailed solution. All steps in the solution are shown, along with an explanation of the relevant geometry theorems and concepts applied. Solutions to many of the multi-step problems begin with a summary of the big-picture strategy to be used in that problem. These solutions are an excellent way to strengthen your geometry knowledge. Understanding why particular strategies and concepts were chosen will make you a much more effective problem solver when you come across similar problems. Note that solving linear equations is an algebraic tool that is used frequently in Regents problems. It is expected that you have mastered that skill, and not all steps in solving linear equations are explicitly explained. For example, if you have the equation $2x + 4 = 12$, then the next line of the solution may read $2x = 8$. It is assumed that you understand that we subtracted 4 from each side in order to separate the variable from the constants in the equation.

- **Self-Analysis Charts** There is a self-analysis chart at the end of each Regents exam. The chart lists the main topic areas and shows which questions fall into which category. Use the chart to record how many points were earned on that practice exam, broken down by each topic area. You can then target areas of weakness for follow-up study and practice. You can also use the charts to find problems covering similar topics on other Regents exams in this book.

About the Exam

TEST LAYOUT AND STANDARDS

The common core geometry Regents is a 3-hour exam and consists of four parts:

Part	Number of Questions	Type of Question	Points per Question	Total Number of Points
I	24	Multiple-choice	2	48
II	7	Constructed-response	2	14
III	3	Constructed-response	4	12
IV	2	Constructed-response	6	12
Total	36	—	—	86

Part I consists of 24 multiple-choice questions worth 2 points each. Part II consists of 7 constructed-response questions worth 2 points each, and part III consists of 3 constructed-response questions worth 4 points each. Part IV consists of 2 constructed-response questions worth 6 points each. One part IV question will either involve a multiple-step proof or ask you to develop an extended logical argument. The second question will require you to use modeling to solve a real-world problem.

The complete set of Geometry Common Core Learning Standards are listed in the Appendix. They are grouped in domains, and each domain accounts for a specified percentage of the total points on the exam, shown in the accompanying table. Note that some domains account for a far greater percentage of points than others. The domains are further divided into clusters with different levels of emphasis on the exam—major, supporting, and additional.

Domain	Cluster	Emphasis	Standard[1]
Congruence (27–34%)	Experiment with transformations in the plane	Supporting	C-CO.1 through G-CO.5
	Understand congruence in terms of rigid motions. Prove geometric theorems	Major	G-CO.6 through G-CO.11
	Make geometric constructions	Supporting	G-CO.12 G-CO.13
Similarity and right triangle trigonometry (29–37%)	Understand similarity in terms of similarity transformations	Major	G-SRT.1 G-SRT.2 G-SRT.3
	Prove theorems involving similarity	Major	G-SRT.4 G-SRT.5
	Define trigonometric ratios and solve problems involving right triangles	Major	G-SRT.6 G-SRT.7 G-SRT.8
Circles (2–8%)	Understand and apply theorems about circles	Additional	G-C.1 G-C.2 G-C.3
	Find arc lengths and areas of sectors of circles	Additional	G-C.5
Expressing geometric properties with equations (12–18%)	Translate between the geometric description and the equation for a conic section	Additional	G.GPE-1
	Use coordinates to prove simple geometric theorems algebraically	Major	G-PE.4 through G-PE.7
Geometric measurements and dimensions (2–8%)	Explain volume formulas and use them to solve problems	Additional	G-GMD.1 G-GMD.3
	Visualize relationships between 2D and 3D objects	Additional	G-GMD.4
Modeling with geometry (8–15%)	Apply geometric concepts in modeling situations	Major	G-MG.1 G-MG.2 G-MG.3

Frequency of Topics

The actual distribution of questions in the last few Regents exams is shown in the following table.

Topic	June 2015	August 2015	June 2016	August 2016	June 2017	August 2017
1. *Angle and Segment Relationships* (define geometric terms and figures, vertical angles, complementary angles, supplementary angles, bisectors and midpoints, parallel and perpendicular lines, proofs of angle and segment relationships)	17, 32			1, 11		
2. *Angle and Segment Relationships in Triangles and Polygons* (angle sum theorem, exterior angle theorem, isosceles triangle theorem, equilateral triangles, interior and exterior angles of polygons, midsegments, centroids, central angles, proofs of triangle and polygon theorems)	34	8, 11	9, 13	4, 8	10, 17	11, 16

Topic	June 2015	August 2015	June 2016	August 2016	June 2017	August 2017
3. *Constructions* (copy a segment, copy an angle; bisect a segment, bisect an angle; perpendicular bisector, perpendicular lines, parallel lines, equilateral triangle, square, regular hexagon, circle inscribed in a triangle, circle circumscribed about a triangle)	25	26	31	28, 32	2	28
4. *Transformations* (definitions and properties of the rigid motions and dilations, preserved properties, compositions of transformations, mapping a figure onto itself with rotations)	2, 4, 10, 16, 33	6, 7, 13, 20, 34	4, 8, 16, 25, 34	2, 9, 27, 29	1, 6, 7, 14, 22, 30, 31, 32	2, 6, 22, 27
5. *Transformations on the Coordinate Plane* (transformations of lines, transformation of points and figures)	18, 22	5, 23, 24	3, 6, 12, 14, 22, 23, 26	5, 26, 33a		

Topic	June 2015	August 2015	June 2016	August 2016	June 2017	August 2017
6. *Triangle Congruence* (SAS, SSS, ASA, AAS, HL criteria; proofs involving triangle congruence; explain the congruence criteria in terms of rigid motions; prove congruence using rigid motions)	24, 30	2, 30, 34	7, 35		9, 33	30, 35
7. *Lines and Segments on the Coordinate Plane* (slope, distance, midpoint, dividing a segment in a specified ratio, equations of lines, determine if a point lies on curve given its equation, area, and perimeter)	3, 9, 27	10, 31	3, 6, 12, 14, 22, 23, 26	14, 15, 16, 18, 21, 30	2, 12, 15, 19	3, 10, 17, 24, 29, 31
8. *Circles on the Coordinate Plane* (equations and graphs of circles, use completing the square to identify the center and radius of a circle)	14	9	3, 6, 12, 14, 22, 23, 26		2, 12, 15, 19	3, 10, 17, 24, 29, 31

Topic	June 2015	August 2015	June 2016	August 2016	June 2017	August 2017
9. *Similarity* (proportional corresponding parts, similar triangle proofs, scaled drawings, perimeter and area of similar figures, proportions in right triangles, prove the Pythagorean theorem)	11, 15, 21, 31, 34	12, 14, 17, 19, 27, 29	2, 5, 17, 21, 27	10, 12, 22	5, 24, 29	5, 7, 9, 18
10. *Trigonometry* (finding a missing side of a right triangle, applications of the sine, cosine, and tangent in word problems, angle of elevation and depression, finding an angle using an inverse trigonometric function, cofunction relationships)	5, 12, 28, 31	4, 32	11, 15, 28, 30	3, 6, 34	3, 13, 21, 36	15, 19, 21

Topic	June 2015	August 2015	June 2016	August 2016	June 2017	August 2017
11. *Parallelograms and Trapezoids* (classifying figures, properties of parallelograms and trapezoids, proofs involving parallelograms and trapezoids, transformations of parallelograms and trapezoids)	13, 26, 33	1, 8, 28, 35	9, 33	7, 24	11, 20	8, 14, 26
12. *Coordinate Geometry Proofs* (proofs on the coordinate plane)	36	22, 33		33b	35	32
13. *Circles* (similarity of circles, central and inscribed angles, angles and segments formed by chords, tangents, and secants, quadrilaterals inscribed in a circle, radian measure, arc length, area of a sector)	8, 20, 29	11, 12, 15, 18	10, 24, 29	19, 23, 25, 35	4, 8, 26	4, 12, 23, 33

Topic	June 2015	August 2015	June 2016	August 2016	June 2017	August 2017
14. *Volume* (prisms, cylinders, cones, pyramids, spheres, properties of solids, calculating volume of solids, justify the formulas for circumference and area of a circle, justify the volume formulas, Cavalieri's principle, identify cross sections and solids of revolution)	1, 6, 23	3, 21	1, 20, 36	3, 13, 20	16, 18, 27, 28	1, 13, 25
15. *Modeling* (modeling real-world situations with solids and planar figures, applications of density, design problems involving area and volume)	7, 19, 35	16, 25, 36	18, 32	17, 36	23, 34	20, 34, 36

HOW THE TEST IS SCORED

All multiple-choice questions are worth 2 points each, and no partial credit is awarded. The constructed-response questions are worth 2, 4, or 6 points each. These are graded according to the scoring rubric issued by the New York State Department of Education. The rubric allows for partial credit if an answer is partially correct.

When determining partial credit, the rubrics often distinguish between *computational* errors and *conceptual* errors. A computational error might be an error in algebra, graphing, or rounding. Computational errors generally will cost you 1 point, regardless of whether the question is a 2-point or 6-point question.

Conceptual errors include using the incorrect formula (volume of a cone instead of a prism) or applying an incorrect relationship (congruent instead of supplementary same side interior angles). Half the credit of the problem is generally deducted for conceptual errors.

Because partial credit is awarded on the constructed-response questions, it is extremely important to show all your work. A correct answer with no work shown will usually cost you most of the points available for that problem.

Your total number of points earned for all four parts will be added to determine your raw score. A conversion table specific for each exam will then be used to convert your raw score to a final scaled score, which is reported to your school. The actual conversion charts are included at the end of each of the Regents exams in this book so you can convert your raw score to a final scaled score. The percentage of points needed to earn a final score of 65% is usually less than 65%, but the effect of the "curve" diminishes as your raw score increases.

CALCULATOR, COMPASS, STRAIGHTEDGE, PEN, AND PENCIL

Graphing calculators are required for the Geometry Common Core exam, and schools must provide one to any students who don't have their own. You may bring your own calculator if you own one, and many students feel more comfortable using their own familiar calculator. Be aware, though, that test administrators may clear the memory and any stored programs you might have saved on your calculator before the exam begins.

Any calculator provided to you will likely have had its memory cleared as well. This process will restore the calculator to its default settings, which may not be the familiar ones you are accustomed to using. The most common setting that may affect your work is the degree versus radian mode. Some calculators have radian mode as the default. It is a good idea to find out what calculator will be provided to you and how to switch between degree and radian mode.

A good working knowledge of the graphing calculator will let you apply more than one method to solving a problem. Some calculator techniques worth knowing are

- Using the graph and intersect feature to solve linear and quadratic equations
- Using tables to find patterns or to quickly apply trial and error to solve a problem
- Finding the square root and cube root of a number
- Using the trigonometric and inverse trigonometric functions

You will also be provided with a compass and straightedge if you do not have your own. Bringing your own compass is highly recommended, since the compasses provided by your school are of unknown quality. Also, it will be less stressful on test day to work with a compass that you are familiar with.

The straightedge provided to you should be used only for making straight lines when graphing or doing constructions. The straightedge may have length markings in inches or centimeters, but these should not be used for determining the length of a segment or checking if two segments are congruent.

You are responsible for bringing your own pen and pencil to the exam. All work must be done in blue or black pen, with the exception of graphs and diagrams, which may be done in pencil.

SCRAP PAPER AND GRAPH PAPER

You are not allowed to bring your own scrap paper or graph paper into the exam. The exam booklet contains one page of scrap paper and one page of graph paper. These are perforated and can be removed from the booklet to make working with them easier. Any work you put on these sheets is *not* graded. If you want a grader to consider any work there, you must copy it into the appropriate space in the booklet.

REFERENCE SHEET

The following reference sheet is provided to you during the Regents exam. The same reference sheet is used for the Algebra I, Geometry, and Algebra II exams. The three sequence and series formulas and the exponential growth and decay formulas are not part of the Common Core Geometry curriculum, so do not be concerned with them. You should be familiar with all the other formulas. Remember, anytime you are asked to calculate an area or volume, check the reference sheet. Many of those formulas are provided.

COMMON CORE HIGH SCHOOL MATH REFERENCE SHEET

(Algebra I, Geometry, Algebra II)

Conversions

1 inch = 2.54 centimeters	1 ton = 2000 pounds
1 meter = 39.37 inches	1 cup = 8 fluid ounces
1 mile = 5280 feet	1 pint = 2 cups
1 mile = 1760 yards	1 quart = 2 pints
1 mile = 1.609 kilometers	1 gallon = 4 quarts
1 kilometer = 0.62 mile	1 gallon = 3.785 liters
1 pound = 16 ounces	1 liter = 0.264 gallon
1 pound = 0.454 kilogram	1 liter = 1000 cubic centimeters
1 kilogram = 2.2 pounds	

Formulas

Triangle	$A = \dfrac{1}{2}bh$
Parallelogram	$A = bh$
Circle	$A = \pi r^2$
Circle	$C = \pi d$ or $C = 2\pi r$
General Prisms	$V = Bh$
Cylinder	$V = \pi r^2 h$
Sphere	$V = \dfrac{4}{3}\pi r^3$
Cone	$V = \dfrac{1}{3}\pi r^2 h$

Pyramid	$V = \frac{1}{3}Bh$
Pythagorean Theorem	$a^2 + b^2 = c^2$
Quadratic Formula	$x = \dfrac{-b \pm \sqrt{b^2 - 4ac}}{2a}$
Arithmetic Sequence	$a_n = a_1 + (n-1)d$
Geometric Sequence	$a_n = a_1 r^{n-1}$
Geometric Series	$S_n = \dfrac{a_1 - a_1 r^n}{1-r}$ where $r \neq 1$
Radians	$1 \text{ radian} = \dfrac{180}{\pi} \text{ degrees}$
Degrees	$1 \text{ degrees} = \dfrac{\pi}{180} \text{ radians}$
Exponential Growth/Decay	$A = A_0 e^{k(t-t_0)} + B_0$

PREPARING FOR THE EXAM

- Preparing to earn a high grade on the Geometry Regents begins well before exam day. Don't try to cram for the Geometry Regents the day before the exam, or even the week before. Successful students begin preparing months ahead of time!

- Make the most of your class time. Take good notes, attempt all homework and classwork, and—most importantly—ask questions when you encounter something you don't understand. Check and correct your work, especially exams and quizzes. Save your exams and quizzes in a folder or binder—they are an excellent resource for review and can help you document your progress.

- You should be working with a good review book throughout the school year. The recommended review book is Barron's *Let's Review: Geometry*. You will be able to start reviewing with that book topic by topic before you reach the end of the curriculum in your geometry class.

- Use the self-analysis charts at the end of each Regents exam in this book to record the number of points you earn within each topic area, and use those results to help identify areas for additional practice.

- Work through at least one Regents exam under "exam conditions"—no distractions, no breaks, and a 3-hour time limit. Measure how much time you spend on each part, and use that as a guide to help establish a time-management plan for the actual exam.

- Review any vocabulary words you are unfamiliar with. Geometry is a vocabulary-intensive subject, and understanding the full meaning and implication of a word encountered in a problem is critical to being able to solve the problem. The glossary at the end of this book is a good resource. One good strategy is to make flash cards with a word on one side and figure on the other.

- Work through as many practice problems as you can. A greater variety of problems that you encounter beforehand means a greater probability of seeing familiar problems on the Regents.

- Become familiar with the formula sheet. You should know what is on the formula sheet and what's not. Also, you should know what each of the symbols and variables represents.

- Read through the step-by-step solutions to the Regents questions for any question you do not get right while practicing. Try to identify why you were not able to correctly answer the question. You will learn from your errors only if you correct them as soon as possible. Be aware of why the errors occur. Common errors are

  – Not remembering a formula or relationship

  – Having difficulty applying multiple steps or concepts in a single problem

  – Having weak algebra skills

  – Having a weak vocabulary

  – Not reading the question carefully

  – Not checking your work

Test-Taking Tips

TIP 1

Be mentally prepared for the exam

SUGGESTIONS

1. Don't try to cram the night before the exam—prepare ahead of time.
2. Walk into the exam room with confidence.
3. You can retake the exam if the grade you earn does not meet the expectation you set for yourself, so don't panic.

TIP 2

Be physically prepared for the exam

SUGGESTIONS

1. Get plenty of sleep the night before, and set an alarm if necessary so you wake up on time.
2. Eat a good breakfast the day of the test.
3. Dress in comfortable clothes.
4. Give yourself plenty of time to get to the test. Late students may be admitted up to a certain time, but they will not be given any additional time at the end of the exam. Check with your teachers for the current policies.

> ### TIP 3
>
> ### Come prepared with all materials

SUGGESTIONS

1. Be sure to have

    - Any identification required by your school
    - Time and room number of your exam
    - Two pens, two pencils, eraser, compass, ruler, and graphing calculator (if you have one)
    - Do **not** bring any restricted items such as cell phones, music or recording devices, food, and drinks. Ask your teacher ahead of time what items are prohibited. If you do need to bring a cell phone or other electronic device to school, plan to leave it in your locker or other secure place. Using a prohibited device during the exam will invalidate your score.

> ### TIP 4
>
> ### Have a consistent plan for working through each problem

SUGGESTIONS

1. Start by reading the question carefully. Common errors to watch out for are

    - Not writing your answer in the correct form, such as a simplified radical or rounded decimal
    - Using the wrong method, such as graphical versus algebraic
    - Solving for a variable but not substituting to find a requested angle or segment measure
    - Missing key words that provide necessary information (parallel, isosceles, regular, "not necessarily true," etc.)

2. Sketch a figure if none is provided. Mark up the figures with relevant dimensions, congruence markings, parallel markings, and so on.

3. Try to break down complex multi-step problems into smaller pieces, looking for relationships that you can apply.

4. For proofs, decide if you are going to write a two-column or paragraph proof.

5. If you don't see how to get to a final answer at first, try applying any formulas or relationships that you recognize. Sometimes finding a certain length or angle may help you see the path to the final answer. Don't panic if you cannot immediately answer a question!

TIP 5

Budget your time

SUGGESTIONS

1. Don't spend too much time on any one problem. If you are having difficulty with a problem, circle it and come back to it at the end. There are going to be a mix of easy, moderate, and difficult questions. You don't want to run out of time before doing all the easy problems.

2. Before the test day, record the time it takes you to do each part of one of the practice Regents exams in this book. Your goal should be to finish all parts in $2\frac{1}{2}$ hours, which will give you 30 minutes to check your work.

TIP 6

Express your answer in the form requested

SUGGESTIONS

1. When required to round to a specified place value, do not round until you get your final answer.

2. When using formulas that involve π, check if the answer is to be "expressed in terms of π" or "rounded" to a certain place value. Expressed in terms of π means leave π out of the calculation, and then put in the π when writing your final answer, for example "32π." If the answer is to be rounded, then you should type π into your calculator along with the appropriate formula. Remember, always use the π button, not 3.14 or any other approximation of π.

3. Simplify radicals when necessary.

4. If your answer to a multiple-choice question is not one of the choices, check to see if it can be rewritten as one of the choices by simplifying a fraction or radical or rearranging an equation. For example, the equation of a line may be written in multiple forms, or an answer may be expressed in terms of π instead of a rounded equivalent.

TIP 7

Maximize your partial credit

SUGGESTIONS

1. Show all work for parts II, III, and IV. Most of the credit is lost if you write a correct answer with no supporting work.

2. Write any formulas used in parts II, III, and IV, and then show each step of the evaluation.

3. Work through the entire problem even if you are not sure if you are correct! If you make an error along the way, but the following work is consistent with the error, you will likely receive partial credit for the correct method.

4. Do not leave any questions blank. There is no penalty for guessing.

TIP 8

Consider alternative methods to solving a problem

SUGGESTIONS

1. Trial and error is a valid method for solving an equation or problem. However, you must clearly write the equation you are using along with at least *three* guesses with an appropriate check. Even if your first guess is correct, demonstrate that you understand the method by showing work for two incorrect guesses and the checks.

2. Some problems can be solved graphically even though a coordinate plane grid is not provided with the problem. You can use the provided scrap graph paper in these cases. The scrap graph paper is not graded, but this is not a concern for multiple-choice problems since work does not need to be shown. For some constructed-response questions, sketching the problem and solution on the graph paper may be a good way to check your result.

TIP 9

Eliminate obviously incorrect choices on multiple-choice questions

SUGGESTIONS

1. You can sometimes rule out choices if you
   - Expect an obtuse or acute angle for an answer
   - Expect a length to be longer or shorter than another known length
2. Use the fact that unless otherwise specified, figures are approximately to scale

TIP 10

Make your paper easy to grade

SUGGESTIONS

1. Work neatly. Your work should be legible and flow from top to bottom.
2. Cross out any incorrect work you do not want the graders to consider on constructed-response questions. You will lose credit if there is both a correct and incorrect answer.
3. Follow directions if you need to change an answer on a multiple-choice bubble sheet. Ask a proctor if you are unsure.

TIP 11

Check your work

SUGGESTIONS

1. If you finish the exam with time left, go back and check as much of your work as you can. After taking the time to prepare for the exam, it would be a waste of that effort to not make use of any extra time you might have.

A Brief Review of Key Geometry Facts and Skills

3.1 ANGLE, LINE, AND PLANE RELATIONSHIPS

NOTATION

Points, segments, lines, and planes are represented using the following notation

Name	Definition	Figure	Notation
Point	A location in space with no length, width, or thickness	•A	Point A
Line	An infinitely long set of points that has no width or thickness	A B M N O m	$\overleftrightarrow{AB}$ $\overleftrightarrow{MN}$, $\overleftrightarrow{NO}$, or $\overleftrightarrow{MO}$ Line m
Ray	A portion of a line with one endpoint and including all points on one side of the endpoint	A B	$\overrightarrow{AB}$

Segment	A portion of a line bounded by two endpoints		$\overline{FG}$
Length of a segment	The distance between the two endpoints of a segment		FG
Plane	A flat set of points with no thickness that extends infinitely in all directions		Plane ABC Plane R
Angle	A figure formed by two rays with a common endpoint		$\angle RST$ or $\angle S$
Angle measure	The amount of opening of an angle, measured in degrees or radians		$\angle 1$ $m\angle 1$

An angle can be classified as a

- **Right angle**—measures 90°
- **Acute angle**—measures less than 90°
- **Obtuse angle**—measures greater than 90° and less than 180°
- **Straight angle**—measures 180°

INTERSECTING, PARALLEL, PERPENDICULAR, AND SKEW

- The intersection of two lines is one point (lines r and s intersecting at point M).
- The intersection of two planes is one line (planes ABC and ABD intersect along $\overleftrightarrow{AB}$).

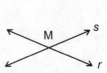 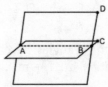

- *Coplanar lines* lie in the same plane, they are either parallel ($\overleftrightarrow{AD}$ and $\overleftrightarrow{BC}$) or intersect ($\overleftrightarrow{AD}$ and $\overleftrightarrow{AB}$).
- *Parallel lines* are coplanar and never intersect. The symbol for parallel is $//$ ($\overleftrightarrow{AD} // \overleftrightarrow{BC}$).
- *Skew lines* are lines that are not coplanar. $\overleftrightarrow{AE}$ and $\overleftrightarrow{GF}$ are skew.
- *Perpendicular lines* intersect at right angles, which measure 90°. The symbol for perpendicular is $\perp$ ($r \perp s$). Line $r \perp$ line s in the accompanying figure. The small square at the point of intersection is a symbol for a right angle. $m\angle ABC + m\angle DEF = 180°$.

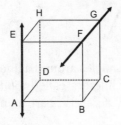

BASIC ANGLE RELATIONSHIPS

- *Supplementary* angles have measures that sum to 180°. $\angle ABC$ and $\angle DEF$ are supplementary.

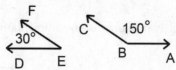

- *Complementary* angles have measures that sum to 90°. $\angle 1$ and $\angle 2$ are complementary. $m\angle 1 + m\angle 2 = 90°$.

- The sum of the measures of adjacent angles around a point equals 360°. $m\angle 1 + m\angle 2 + m\angle 3 + m\angle 4 = 360°$.

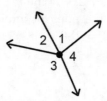

- The sum of the measures of adjacent angles around a line equals 180°. Two adjacent angles that form a straight line are called a *linear pair*, and are supplementary. $m\angle 1 + m\angle 2 = 180°$. $\angle 1$ and $\angle 2$ are a linear pair.

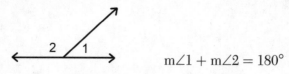

$$m\angle 1 + m\angle 2 = 180°$$

- The measure of a whole equals the sum of the measures of its parts. $m\angle ABC = m\angle 1 + m\angle 2 + m\angle 3$.

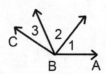

- *Vertical angles* are the congruent opposite angles formed by intersecting lines. ∠1 and ∠3 are vertical angles; therefore, ∠1 ≅ ∠3. ∠2 and ∠4 are vertical angles; therefore, ∠2 ≅ ∠4.

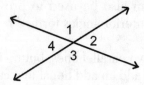

- *Angle bisectors* divide angles into two congruent angles. $\overrightarrow{CD}$ bisects ∠ACB, ∠ACD ≅ ∠BCD.

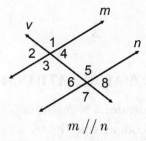

ANGLES FORMED BY PARALLEL LINES

When two parallel lines (lines m and n) are intersected by a transversal (line v), eight angles are formed. Any pair of these angles will be either supplementary or congruent. Certain pairs are given special names. The named pairs have the following properties:

$$m \,/\!/\, n$$

Alternate interior angles are congruent. Alternate interior angles lie between parallel lines m and n on opposite sides of transversal v.

Same side interior angles are supplementary. Same side interior angles lie between parallel lines m and n on the same side of transversal v. ∠4 and ∠5 are supplementary, ∠3 and ∠6 are supplementary. m∠4 + m∠5 = 180°, m∠3 + m∠6 = 180°.

Corresponding angles are congruent. Corresponding angles are the interior/exterior pair of congruent angles on the same side of the transversal. $\angle 4 \cong \angle 8$, $\angle 1 \cong \angle 5$, $\angle 2 \cong \angle 6$, $\angle 3 \cong \angle 7$.

These relationships can also be used to prove two lines are parallel. For example, if the alternate interior angles formed by two lines are congruent, then the lines are parallel.

If a problem involves two parallel lines, but no transversal that intersects both lines, it may be helpful to add an additional line called an auxiliary line. In order to find m$\angle BED$ in the accompanying figure, the auxiliary line $\overleftrightarrow{FEG}$ is drawn parallel to $\overline{CD}$ and $\overline{AB}$. Now the two alternate interior angle relationships can be used:

$$\text{m}\angle FED = \text{m}\angle CDE = 38° \qquad \text{alternate interior angles}$$
$$\text{m}\angle FEB = \text{m}\angle ABE = 32° \qquad \text{alternate interior angles}$$
$$\text{m}\angle DEB = \text{m}\angle FED + \text{m}\angle FEB \quad \text{angle addition}$$
$$= 38° + 32°$$
$$= 70°$$

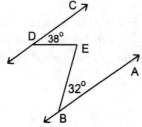
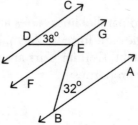

BASIC LINE/SEGMENT/RAY RELATIONSHIPS

In segments, the point that divides a segment into two congruent segments is called a *midpoint*. In the accompanying figure, S is the midpoint of $\overline{RT}$, and $\overline{RS} \cong \overline{ST}$. Any line, segment, or ray that intersects a segment at its midpoint is called a *segment bisector*, or simply *bisector*. Line m bisects $\overline{RT}$ at its midpoint S. Lines can bisect segments, but lines cannot be bisected themselves because they do not have a finite length.

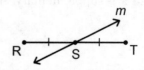

- A line, segment, or ray that is perpendicular to and bisects a segment is called a perpendicular bisector. $\overleftrightarrow{CD}$ is the perpendicular bisector of $\overline{AB}$. The four angles formed by their intersection are right angles, and $\overline{AE} \cong \overline{BE}$.

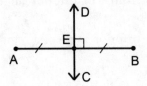

As with angles, the length of a divided segment is equal to the sum of its parts. In the accompanying figure, $MN + NO = MO$. If point N is also a midpoint, then the additional relationship $MN = NO$ would be true.

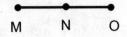

ANGLES IN POLYGONS

Definitions:

- **Polygon**—a closed planar figure with straight sides
- **Regular polygon**—a polygon whose sides and angles are all congruent
- **Interior angle**—the angle inside the polygon formed by two adjacent sides
- **Exterior angle**—the angle formed by a side and the extension of an adjacent side in a polygon
- **Triangle**—a polygon with 3 sides
- **Quadrilateral**—a polygon with 4 sides
- **Pentagon, hexagon, octagon, decagon**—polygons with 5, 6, 8, and 10 sides

In the accompanying pentagon, angles 1 through 5 are interior angles and angles 6 through 10 are exterior angles.

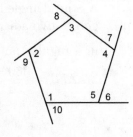

The following relationships can be used to find interior and exterior angles of a polygon:

- The sum of the measures of interior angles of a polygon with n sides equals $180° \times (n - 2)$.

- The measure of one interior angle of a regular polygon with n sides equals $\dfrac{180°(n-2)}{n}$.

- The sum of the measures of exterior angles of a polygon with n sides equals $360°$.

- The measure of one exterior angle of a regular polygon with n sides equals $\dfrac{360°}{n}$.

Using the accompanying figure of the pentagon on page 26:

$$m\angle 1 + m\angle 2 + m\angle 3 + m\angle 4 + m\angle 5 = 180 \times (n-2).$$
$$= 180(5-2)$$
$$= 540°$$
$$m\angle 6 + m\angle 7 + m\angle 8 + m\angle 9 + m\angle 10 = 360°$$

If the pentagon is regular,

$$m\angle 1 = \frac{180°(n-2)}{n}$$
$$= \frac{180°(5-2)}{5}$$
$$= 108°$$
$$m\angle 6 = \frac{360°}{n}$$
$$= \frac{360°}{5}$$
$$= 72°$$

- A central angle is the angle formed by connecting the center of the polygon to two adjacent vertices, as shown. The measure of a central angle in a regular polygon with n sides equals $\dfrac{360°}{n}$. For pentagon $ABCDE$, $m\angle APB = \dfrac{360°}{5} = 72°$.

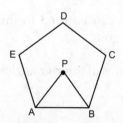

Practice Exercises

1 In the figure of the rectangular prism, which of the following is true?

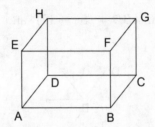

- (1) Points E, H, D, and A are coplanar and collinear.
- (2) $\overline{HD}$ is skew to $\overline{CD}$, and $\overline{CD} \perp \overline{CG}$.
- (3) $\overline{EA} \mathbin{/\!/} \overline{CG}$, and $\overline{EH}$ is skew to $\overline{FB}$.
- (4) $\overline{EA} \perp \overline{BC}$, and $\overline{AB} \mathbin{/\!/} \overline{CD}$.

2 Which of the following would be the best definition of an angle?
- (1) the union of two rays with a common endpoint
- (2) a geometric figure measured in degrees
- (3) one-third of a triangle
- (4) a line that bends

3 $\overleftrightarrow{BFA} \perp \overrightarrow{CF}$ and m$\angle CFE = 42°$. Find m$\angle BFD$.

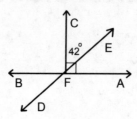

(1) 42°

(2) 45°

(3) 48°

(4) 52°

4 $\overrightarrow{BC}$ bisects $\angle ABD$. If m$\angle ABD = (8x - 12)°$ and m$\angle ABC = (3x + 4)°$, find m$\angle ABD$.

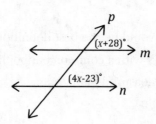

(1) 10°

(2) 20°

(3) 34°

(4) 68°

5 Line p intersects lines m and n. For what values of x could you conclude that m is parallel to n?

(1) 12°

(2) 17°

(3) 35°

(4) 62°

6 Points F, G, H, and I are collinear. G is the midpoint of $\overline{FH}$, H is the midpoint of $\overline{GI}$, $\overline{FG} = (x^2 + 5x)$, and $\overline{HI} = 3x + 8$.

What is the length of $\overline{FG}$?

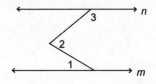

(1) 14   (3) 24

(2) 20   (4) 32

7 In the accompanying figure, lines m and n are parallel. If $m\angle 1 = 32°$ and $m\angle 2 = 78°$, what is the measure of $\angle 3$?

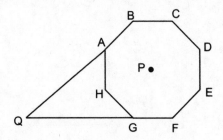

(1) 102°   (3) 134°

(2) 110°   (4) 148°

8 A regular polygon has interior angles that each measure 156°. How many sides does the polygon have?

9 Regular octagon $ABCDEFGH$ is shown in the figure below.

Sides $\overline{AB}$ and $\overline{FG}$ are both extended to Q.
What is the measure of $\angle Q$?

(1) 22.5°   (3) 45°

(2) 30°   (4) 60°

10 In regular hexagon *ABCDEF*, $\overline{AD}$ and $\overline{FC}$ intersect at *O*. What is the measure of ∠*AOF*?

11 In the accompanying figure, $\overline{WPX}$ intersects $\overline{YPZ}$ at *P*.

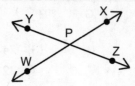

Prove the theorem that the opposite angles formed by intersecting lines are congruent, ∠*WPY* ≅ ∠*ZPX*.

Solutions

1 Choice (1) is not correct since E, H, and D do not lie on a single straight line and are not collinear.

Choice (2) is not correct since $\overline{HD}$ and $\overline{CD}$ are perpendicular, so they cannot be skew.

Choice (4) is not correct since $\overline{EA}$ and $\overline{BC}$ do not intersect, so they cannot be perpendicular.

The correct choice is (**3**).

2 Considering each choice:

(1) This definition is precise and uses previously defined geometric terms.

(2) This definition is not precise. A counterexample would be an arc, which can also have a degree measure.

(3) Not all angles are part of triangles.

(4) This definition is not precise, since "bend" is not a defined geometric term.

The correct choice is (**1**).

3 $\text{m}\angle CFE + \text{m}\angle EFA = 90°$ Perpendicular lines form $90°$ angles.

$\quad\quad 42° + \text{m}\angle EFA = 90°$

$\quad\quad\quad\quad\quad \text{m}\angle EFA = 48°$

$\quad\quad\quad\quad\quad \text{m}\angle BFD = \text{m}\angle EFA$ Vertical angles are congruent

$\quad\quad\quad\quad\quad \text{m}\angle BFD = 48°$

The correct choice is (**3**).

4 $m\angle ABD = m\angle ABC + m\angle CBD$ angle addition

   $m\angle ABC = m\angle CBD$ bisector forms

                                $2 \cong$ segments

   $m\angle ABC + m\angle ABC = m\angle ABD$ substitute $m\angle ABC$ for

                                  $m\angle CBD$

$$3x + 4 + 3x + 4 = 8x - 12$$
$$6x + 8 = 8x - 12$$
$$x = 10$$

$$m\angle ABD = (8 \cdot 10 - 12)° \quad \text{evaluate } m\angle ABD \text{ for}$$
$$x = 10$$
$$= 68°$$

The correct choice is (**4**).

5 The two angles indicated in the figure are corresponding angles, so lines m and n are parallel if the corresponding angles are congruent, or have the same angle measure.

$$4x - 23 = x + 28$$
$$3x = 51$$
$$x = 17$$

The correct choice is (**2**).

6 A bisector divides a segment into two congruent halves, so $FG = GH$ and $GH = HI$. Combine these relationships to obtain $FG = HI$.

$$x^2 + 5x = 3x + 8$$
$$x^2 + 2x - 8 = 0$$
$$(x - 2)(x + 4) = 0 \qquad \text{factor}$$
$$(x - 2) = 0 \qquad (x + 4) = 0 \qquad \text{zero product property}$$
$$x = 2 \qquad x = -4$$

Eliminate the solution $x = -4$, because that would lead to a negative or zero length; therefore, $x = 2$.

$$FG = x^2 + 5x$$
$$= 2^2 + 5 \cdot 2$$
$$= 14$$

The correct choice is **(1)**.

7 Sketch the auxiliary line parallel to m and n as shown.

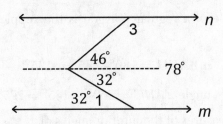

The 78° angle is divided into a lower and upper part. The lower part is an alternate interior angle to $\angle 1$, so it must measure 32°. The measure of the upper part is equal to $78° - 32° = 46°$.

$\angle 3$ and the 46° angle are same side interior angles, so they must be supplementary.

$$m\angle 3 + 46° = 180°$$
$$m\angle 3 = 134°$$

The correct choice is **(3)**.

8 Each interior angle of a polygon measures $\dfrac{180°(n-2)}{n}$ degrees.

$$\frac{180°(n-2)}{n} = 156$$
$$180(n-2) = 156n$$
$$180n - 360 = 156n$$
$$24n = 360$$
$$n = 15$$

The polygon has 15 sides.

9 The sum of the exterior angles of a polygon equals 360°, and in a regular polygon each exterior angle measures $\dfrac{360°}{n}$.

The strategy is to find each interior angle of quadrilateral $QAHG$.

$$m\angle QAH = m\angle QGH = \frac{360°}{8}$$
$$= 45°$$

Each interior angle of a regular polygon is the supplement of an exterior angle.

$$m\angle AHG = 180° - 45° = 135°$$

The exterior convex angle $\angle AHG = 360° - 135° = 225°$.

The sum of the measures of the interior angles of

$$QAHG = (n-2) \cdot 180°$$
$$= (4-2) \cdot 180°$$
$$= 360°$$

$$m\angle Q + m\angle QAH + m\angle AHG + m\angle QGH = 360°$$
$$m\angle Q + 45° + 225° + 45° = 360°$$
$$m\angle Q + 315° = 360°$$
$$m\angle Q = 45°$$

The correct choice is **(3)**.

10 The intersection of the two diagonals $\overline{FC}$ and $\overline{AD}$ is the center of the regular polygon, so $\angle AOF$ is a central angle.

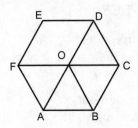

$$\begin{aligned}
\text{m}\angle AOF &= \frac{360°}{n} \quad \text{central angle formula for a regular polygon} \\
&= \frac{360°}{6} \\
&= 60°
\end{aligned}$$

11 The strategy is to show $\angle WPY$ and $\angle ZPX$ are supplementary to the same angle.

Statement	Reason
1. $\overleftrightarrow{WPX}$ intersects $\overleftrightarrow{YPZ}$ at P.	1. Given
2. $\angle WPY$ and $\angle YPX$ are a linear pair. $\angle ZPX$ and $\angle YPX$ are a linear pair.	2 Definition of a linear pair
3. $\angle WPY$ and $\angle YPX$ are supplementary. $\angle ZPX$ and $\angle YPX$ are supplementary.	3. Linear pairs are supplementary
4. $\angle WPY \cong \angle ZPX$	4. Angles supplementary to the same angle are congruent

3.2 TRIANGLE RELATIONSHIPS

CLASSIFYING TRIANGLES

A triangle is a polygon with three sides.
   Triangles can be classified by their sides or by their angles.

- By Sides
   Scalene—no sides are congruent.
   Isosceles—two sides are congruent.
   Equilateral—three sides are congruent.

- By Angle
   Acute—all angles are acute.
   Right—one angle is a right angle.
   Obtuse—one angle is obtuse.

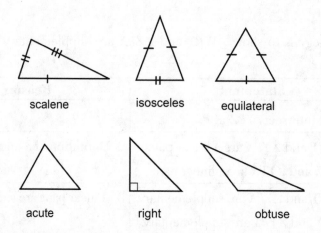

SPECIAL SEGMENTS AND POINTS OF CONCURRENCY IN TRIANGLES

There are four special segments that can be drawn in a triangle, and every triangle has three of each. These are the altitude, median, angle bisector, and perpendicular bisector, shown in the accompanying figure.
   Altitude—a segment from a vertex perpendicular to the opposite side
   Median—a segment from a vertex to the midpoint of the opposite side
   Angle bisector—a line, segment, or ray passing through the vertex of a triangle and bisecting that angle

Perpendicular bisector—a line, segment, or ray that is perpendicular to and passes through the midpoint of a side

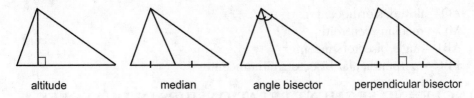

| altitude | median | angle bisector | perpendicular bisector |

Special segments in triangles

The three medians of a triangle will all intersect at a single point called the centroid. We say the medians are *concurrent* at the centroid, or the centroid is the *point of concurrency*. The other special segments will be concurrent as well. The following table illustrates each of the points of concurrency.

Segment	Concurrent at	Feature
Perpendicular bisectors	Circumcenter	Center of the circumscribed circle, equidistant from each of the three vertices
Angle bisectors	Incenter	Center of the inscribed circle, equidistant from the three sides of the triangle
Medians	Centroid	Divides each median in 2:1 ratio and is the center of gravity of the triangle
Altitudes	Orthocenter	Located inside acute triangles, on a vertex of right triangles, and outside obtuse triangles

The following phrase can help you remember the points of concurrency:

"**All** **of** **m**y **c**hildren **are** **b**ringing **in** **p**eanut **b**utter **c**ookies"

AO—altitudes/orthocenter
MC—medians/centroid
ABI—angle bisectors/incenter
PBC—perpendicular bisectors/circumcenter

ANGLE AND SEGMENT RELATIONSHIPS IN TRIANGLES

Angle sum theorem	The sum of the measures of the interior angles of a triangle equals 180°.	$m\angle A + m\angle B + m\angle C = 180°$
Exterior angle theorem	The measure of any exterior angle of a triangle equals the sum of the measures of the nonadjacent interior angles.	$m\angle 1 = m\angle 3 + m\angle 4$
Isosceles triangle theorem and its converse	If two sides of a triangle are congruent, then the angles opposite them are congruent. If two angles in a triangle are congruent, then the sides opposite them are congruent.	$\angle A \cong \angle B, \overline{AC} \cong \overline{BC}$

Equilateral triangle theorem	All interior angles of an equilateral triangle measure 60°.	$\overline{AB} \cong \overline{BC} \cong \overline{AC},$ $m\angle A = m\angle B = m\angle C = 60°$
Pythagorean theorem	In a right triangle, the sum of the squares of the legs equals the square of the hypotenuse.	$a^2 + b^2 = c^2$

Practice Exercises

1 In $\triangle FGH$, K is the midpoint of $\overline{GH}$. What type of segment is $\overline{FK}$?

(1) median

(2) altitude

(3) angle bisector

(4) perpendicular bisector

2 The side lengths of a triangle are 6, 8, and 10.

The triangle can be classified as

(1) equilateral

(2) acute

(3) obtuse

(4) right

3 The angle measures of a triangle are $(6x + 6)°$, $(8x − 8)°$, and $(12x)°$. The triangle can be classified as

(1) obtuse

(2) equilateral

(3) isosceles

(4) scalene

4 Line m // line n, and $\triangle ABC$ is isosceles with $AC = BC$. What is the measure of $\angle 3$?

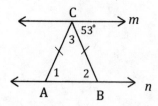

(1) 53°

(2) 60°

(3) 63.5°

(4) 74°

5 In $\triangle ABC$, $\overline{CY}$ is an angle bisector, m$\angle AYC = 71°$ and m$\angle B = 23°$. What is the measure of angle A?

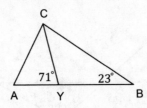

(1) 61°   (3) 76°

(2) 71°   (4) 86°

6 $\triangle RST$ is a right triangle with right angle $\angle RST$. $\triangle PSR$ is equilateral. Find the measure of angle T.

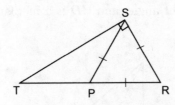

(1) 15°   (3) 30°

(2) 25°   (4) 45°

7 In the figure below, triangle ABC is isosceles with $AC = BC$, m$\angle A = 70°$, and $\overline{AC} \parallel \overrightarrow{BE}$. Find the measure of $\angle CBE$.

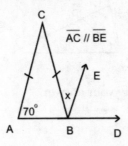

(1) 20°   (3) 35°

(2) 25°   (4) 40°

8 $\overline{JHFK}$ intersects $\overline{IHG}$ at H. If m$\angle KFG = 156°$, m$\angle G = 117°$, m$\angle JHI = (2x + 15)°$, and m$\angle J = (4x + 6)°$, what is the measure of angle J?

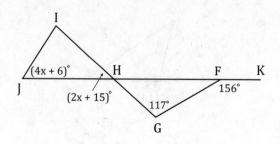

(1) 42°   (3) 62°

(2) 54°   (4) 70°

9 $\angle A$ of $\triangle ABC$ is a right angle, and $\overline{CD}$ is a median. Find the length CD if $AC = 9$ and $BC = 13$.

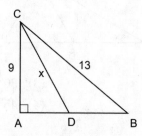

10 In $\triangle ABC$ shown below, m$\angle B = 36°$, $AB = BC$, and $\overline{AD}$ bisects $\angle BAC$.

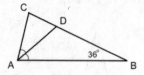

Is $\triangle CAD$ isosceles? Justify your answer.

11 $\overline{RTV}$ intersects $\overline{STU}$ at T.

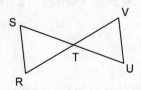

What is the relationship between the sum (m∠R + m∠S) and the sum (m∠V + m∠U)? Explain your reasoning.

Solutions

1 A median is a segment that joins a vertex of a triangle to the midpoint of the opposite side.

The correct choice is (**1**).

2 The three side lengths satisfy the Pythagorean theorem:

$$a^2 + b^2 = c^2$$
$$6^2 + 8^2 = 10^2$$
$$36 + 64 = 100$$
$$100 = 100$$

Therefore, the triangle is a right triangle.

The correct choice is (**4**).

3 $6x + 6 + 8x - 8 + 12x = 180$ angle sum theorem

$$26x - 2 = 180$$
$$26x = 182$$
$$x = 7$$

Substitute $x = 7$ to find the three angle measures.

$$(6 \cdot 7 + 6)^\circ = 48^\circ$$
$$(8 \cdot 7 - 8)^\circ = 48^\circ$$
$$(12 \cdot 7)^\circ = 84^\circ$$

Since the triangle has two congruent angles, it must have two congruent sides, so it is isosceles.

The correct choice is (**3**).

4   $m\angle 2 = 53°$ alternate interior angles
formed by parallel lines
are congruent

$m\angle 1 = m\angle 2$ angles opposite congruent
sides in a triangle are
congruent

$m\angle 1 = 53°$

$$m\angle 1 + m\angle 2 + m\angle 3 = 180°\quad \text{angle sum theorem}$$
$$53° + 53° + m\angle 3 = 180°$$
$$106° + m\angle 3 = 180°$$
$$m\angle 3 = 74°$$

The correct choice is **(4)**.

5
$$m\angle AYC = m\angle B + m\angle BCY\quad \text{exterior angle}$$
theorem in $\triangle YBC$
$$71° = 23° + m\angle BCY$$
$$m\angle BCY = 48°$$
$$m\angle ACY = m\angle BCY$$

angle bisector forms
2 congruent angles

$$m\angle ACY = 48°$$
$$m\angle A + m\angle AYC + m\angle ACY = 180°$$

angle sum theorem
in $\triangle AYC$

$$m\angle A + 71° + 48° = 180°$$
$$m\angle A + 119° = 180°$$
$$m\angle A = 61°$$

The correct choice is **(1)**.

6    $m\angle R = m\angle RPS = m\angle PSR = 60°$ equilateral
                                                      triangle theorem

   $m\angle RST = 90°$ right angles
                                                      measure 90°

   $m\angle T + m\angle RST + m\angle R = 180°$ angle sum theorem
                                                      in $\triangle RST$

$$m\angle T + 90° + 60° = 180°$$
$$m\angle T + 150° = 180°$$
$$m\angle T = 30°$$

The correct choice is **(3)**.

7 The strategy is to first find $m\angle C$, which must be congruent to $\angle CBE$.

$$m\angle ABC = m\angle A = 70°$$ isosceles triangle
                                     theorem

$$m\angle C + m\angle A + m\angle ABC = 180°$$ triangle angle sum
                                                  theorem

$$m\angle C + 70° + 70° = 180°$$
$$m\angle C + 140° = 180°$$
$$m\angle C = 40°$$

$$m\angle CBE = m\angle C$$ congruent alternate
                               interior angles

$$m\angle CBE = 40°$$

The correct choice is **(4)**.

8 Start by finding m$\angle GHK$ by applying the exterior angle theorem in $\triangle FGH$.

$$m\angle GHF + m\angle FGH = m\angle KFG \qquad \text{exterior angle theorem}$$
$$m\angle GHF + 117° = 156°$$
$$m\angle GHF = 39°$$
$$m\angle JHI = m\angle GHF \qquad \text{vertical angles are}$$
$$\text{congruent}$$
$$(2x + 15)° = 39°$$
$$2x = 24$$
$$x = 12$$
$$m\angle J = (4 \cdot 12 + 6)° \quad \text{substitute } x = 12$$
$$= 54°$$

The correct choice is (**2**).

9 Start by using the Pythagorean theorem in $\triangle ABC$ to find AB. From AB, calculate AD, and then use the Pythagorean theorem again on $\triangle ADC$ to find CD.

$$a^2 + b^2 = c^2 \qquad \text{Pythagorean theorem}$$
$$AB^2 + AC^2 = BC^2$$
$$AB^2 + 9^2 = 13^2$$
$$AB^2 + 81 = 169$$
$$AB^2 = 88$$
$$AB = \sqrt{88}$$
$$= \sqrt{4 \cdot 22}$$
$$= 2\sqrt{22}$$

Median $\overline{CD}$ intersects midpoint D, so

$$AD = \frac{1}{2}AB$$
$$AD = \frac{1}{2}\left(2\sqrt{22}\right)$$
$$= \sqrt{22}$$

Now apply the Pythagorean theorem in $\triangle ADC$:

$$a^2 + b^2 = c^2$$
$$AD^2 + AC^2 = CD^2$$
$$\left(\sqrt{22}\right)^2 + 9^2 = CD^2$$
$$22 + 81 = CD^2$$
$$103 = CD^2$$
$$CD = \sqrt{103}$$

10 $\triangle BAC$ has $2 \cong$ sides $\overline{AB}$ and $\overline{BC}$, so it is isosceles. The strategy is to find the measures of $\angle C$, $\angle ADC$, and $\angle CAD$.

$$\text{m}\angle BAC = \text{m}\angle C \quad \text{isosceles triangle theorem in } \triangle ABC$$

$$\text{m}\angle BAC + \text{m}\angle C + \text{m}\angle B = 180° \quad \text{angle sum theorem in } \triangle ABC$$

$$\text{m}\angle BAC + \angle BAC + \text{m}\angle B = 180° \quad \text{substitute m}\angle BAC \text{ for m}\angle C$$

$$2\text{m}\angle BAC + 36° = 180°$$
$$2\text{m}\angle BAC = 144°$$
$$\text{m}\angle BAC = 72°$$
$$\text{m}\angle C = 72°$$

From the angle bisector, we know

$$m\angle CAD = \frac{1}{2}m\angle BAC$$
$$= \frac{1}{2}(72°)$$
$$= 36°$$

Find m$\angle CDA$ by applying the angle sum theorem to $\triangle CAD$.

$$m\angle CDA + m\angle C + m\angle CAD = 180°$$
$$m\angle CDA + 72° + 36° = 180°$$
$$m\angle CDA + 108° = 180°$$
$$m\angle CDA = 72°$$

Both $\angle C$ and $\angle CDA$ measure 72°; therefore, $\overline{AC} \cong \overline{AD}$, and $\triangle CAD$ is isosceles.

11 m$\angle RTS$ + m$\angle R$ + m$\angle S = 180°$ from the angle sum theorem. Also, m$\angle VTU$ + m$\angle V$ + m$\angle U = 180°$ for the same reason; therefore,

$$m\angle RTS + m\angle R + m\angle S = m\angle VTU + m\angle V + m\angle U$$

$\angle RTS$ and $\angle VTU$ are congruent vertical angles, so we can subtract their measure from each side of the equation, resulting in

$$m\angle R + m\angle S = m\angle V + m\angle U$$

Therefore, the two sums are equal.

3.3 CONSTRUCTIONS

COPY AN ANGLE

- Given angle $\angle ABC$ and $\overrightarrow{DE}$, construct $\angle EDF$ congruent to $\angle ABC$.
    1. Place point on B and make an arc intersecting the angle at R and S.
    2. With the same compass opening, place point on E and make an arc intersecting at T.
    3. Place point on R and pencil on S and make a small arc.
    4. With the same compass opening, place point on T and make an arc intersecting the previous one at F.
    5. $\angle DEF$ is congruent to $\angle ABC$.

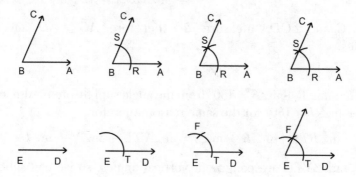

EQUILATERAL TRIANGLE

- Given segment $\overline{AB}$, construct an equilateral triangle with side length AB.
    1. Place point on A and pencil on B and make a quarter circle.
    2. Place point on B and pencil on A and make a quarter circle that intersects the first circle at C.
    3. $\triangle ABC$ is an equilateral triangle.

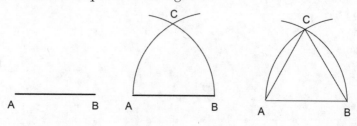

ANGLE BISECTOR

- Given angle $\angle ABC$, construct angle bisector $\overrightarrow{BD}$.
    1. Place point on B and make an arc intersecting $\overrightarrow{BA}$ and $\overrightarrow{BC}$ at R and S.
    2. Place point on R and make an arc in interior of the angle.
    3. With the same compass opening, place point on S and make an arc that intersects the previous arc at point D.
    4. $\overrightarrow{BD}$ is the angle bisector.

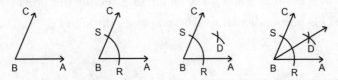

PERPENDICULAR BISECTOR

- Given segment $\overline{AB}$, construct the perpendicular bisector of $\overline{AB}$.
    1. With compass open more than half the length of $\overline{AB}$, place the point at A and make a semicircle running above and below $\overline{AB}$.

    With the same compass opening, place the point at B and make a semicircle running above and below $\overline{AB}$ so that it intersects the first semicircle at R and S.
    2. $\overleftrightarrow{RS}$ is the perpendicular bisector of $\overline{AB}$.

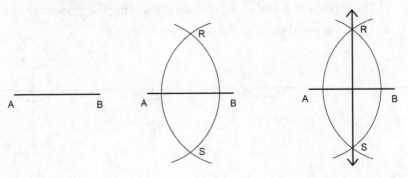

PERPENDICULAR TO LINE FROM A POINT NOT ON THE LINE

- Given line $\overleftrightarrow{AB}$ and point P not on $\overleftrightarrow{AB}$, construct a line perpendicular to $\overleftrightarrow{AB}$ passing through P.

    1. Place point at P and make an arc intersecting $\overleftrightarrow{AB}$ at R and S (extend $\overleftrightarrow{AB}$ if necessary).
    2. Place point at R and make an arc on opposite side of line as P.
    3. Place point at S and make an arc intersecting the previous arc at Q.
    4. $\overleftrightarrow{PQ}$ is perpendicular to $\overleftrightarrow{AB}$ and passes through P.

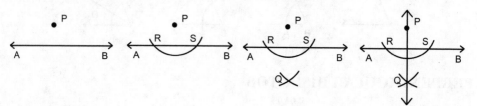

PERPENDICULAR TO LINE FROM A POINT ON THE LINE

- Given line $\overleftrightarrow{AB}$ and point P on $\overleftrightarrow{AB}$ between A and B, construct a line perpendicular to $\overleftrightarrow{AB}$ and passing through P.

    1. Place point at P and make an arc intersecting $\overleftrightarrow{AB}$ at R and S.
    2. Place point at R and make an arc.
    3. Place point at S and make an arc intersecting the previous arc at Q.
    4. $\overleftrightarrow{PQ}$ is perpendicular to $\overleftrightarrow{AB}$ and passes through P.

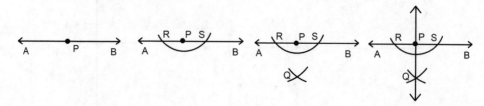

PARALLEL TO LINE THROUGH POINT OFF THE LINE

- Given $\overleftrightarrow{AB}$ and point P not on $\overleftrightarrow{AB}$, construct a line parallel to $\overleftrightarrow{AB}$ and passing through P.

  1. Construct a line passing through P and intersecting $\overleftrightarrow{AB}$ at R.

  2. With point at R, make an arc intersecting $\overleftrightarrow{PR}$ and $\overleftrightarrow{AB}$ at S and T.

  3. With point at P and the same compass opening, make an arc intersecting $\overleftrightarrow{PR}$ at U.

  4. With point at T, make an arc intersecting at S.

  5. With point at U and the same compass opening, make an arc intersecting at V.

  6. $\overleftrightarrow{PV}$ is parallel to $\overleftrightarrow{AB}$ and passes through point P.

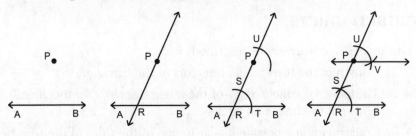

INSCRIBE A REGULAR HEXAGON OR EQUILATERAL TRIANGLE IN A CIRCLE

  1. Construct circle P of radius PA.

  2. Using the same radius as the circle, place point on A and make an arc intersecting circle at B.

  3. Place point at B and make an arc intersecting circle at C. Continue making arcs intersecting at D, E, and F.

  4. Connect each point to form regular hexagon $ABCDEF$ or every other point to form equilateral triangle ACE.

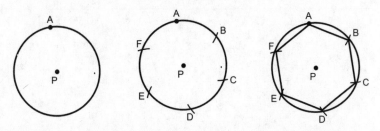

INSCRIBE A SQUARE IN A CIRCLE

- Given circle T
    1. Construct a diameter through the center point T.
    2. Construct the perpendicular bisector of the first diameter, which will also be a diameter.
    3. The intersections of the diameters with the circle are the vertices of the square.

INSCRIBED CIRCLE

- Given $\triangle ABC$, construct the inscribed circle.
    1. Construct the three angle bisectors of the triangle.
    2. The point of concurrency of the angle bisectors, or incenter I, is the center of the inscribed circle.
    3. Construct a line perpendicular to one of the sides of the triangle that passes through the incenter I. Label the point of intersection with the side of the triangle P.
    4. Construct the inscribed circle with center I and radius IP.

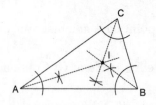 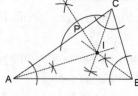 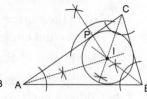

CIRCUMSCRIBED CIRCLE

- Given △ABC, construct the circumscribed circle.

  1. Construct the three perpendicular bisectors of the triangle.

  2. The point of concurrency of the perpendicular bisectors, or circumcenter P, is the center of the circumscribed circle.

  3. Construct the circumscribed circle with center P and radius PA. (The circle should pass through points B and C as well.)

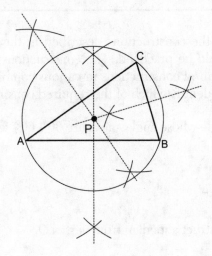

Practice Exercises

Directions: Each of the constructions described in this chapter is a required construction and should be practiced. The constructions in this set of practice problems use the required constructions in various combinations. The first step in each problem is to identify which of the required constructions is called for.

1 Construct a triangle whose angles measure 30°, 60°, and 90°.

2 Given △*JOT*, construct a median from vertex *O*.

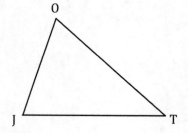

3 Construct the centroid of △*HOT* using a compass and straightedge. Use the compass to confirm your construction by demonstrating the 2:1 ratio. Leave all marks of construction.

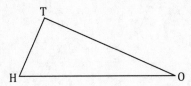

4 Given △*XYZ*, construct the altitude from vertex *X*.

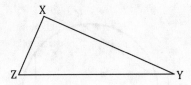

5 Given △*XYZ*, construct a midsegment that intersects sides $\overline{XZ}$ and $\overline{YZ}$.

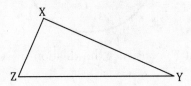

6 Given $\overline{AB}$ and $\overline{BC}$, construct parallelogram *ABCD*.

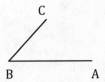

7 Construct a triangle whose angles measure 45°, 45°, and 90° and has $\overline{AB}$ as one of its legs.

A   B

8 Construct a 15° angle.

9 $\overline{JK}$ and $\overline{LM}$ are chords in circle P. Construct the location of point P.

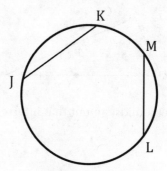

10 Given $\overline{AB}$ and point P, construct the dilation of $\overline{AB}$ with a scale factor of 2 and center at P.

11 Construct the translation of point P by vector $\overrightarrow{AB}$.

12 P' is the reflection of P over line m. Construct line m.

P'

P

13 Construct point Q', the rotation of point Q about center P by $\angle ABC$.

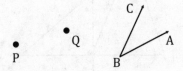

Solutions

1 Construct an equilateral triangle, then bisect one of the angles.

2 Construct A, the midpoint of $\overline{JT}$, using the "perpendicular bisector" construction. Connect O to A. $\overline{OA}$ is a median of $\triangle JOT$.

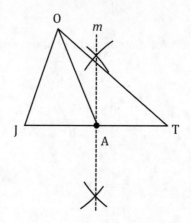

3 The centroid is the point of concurrency of the medians.

- First construct midpoints of two sides of the triangle.
- Then use those midpoints to construct the medians.
- The centroid, G, is the point where the medians intersect.

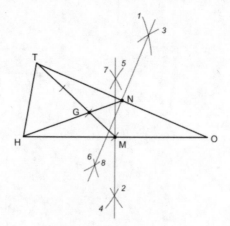

To confirm the centroid divides a median in a 2:1 ratio, measure the distance *GM* and show the distance *TG* is twice *GM*:

- Place the compass point at *M* and make an arc at *G*.
- With the same compass opening, place the point at *G* and make a 2nd arc in the direction of *T*.
- With the same compass opening, you should be able to make a 3rd arc starting at the 2nd and ending at *T*.

4 Using the "perpendicular to line from point not on the line" construction, construct a line perpendicular to $\overline{YZ}$ through point *X*. Let *A* be the intersection of the perpendicular line with $\overline{YZ}$. $\overline{AX}$ is the altitude of △*XYZ*.

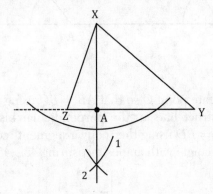

5 Using the "perpendicular bisector" construction, construct the midpoints of $\overline{XZ}$ and $\overline{YZ}$. Label these *A* and *B*. $\overline{AB}$ is a midsegment of △*XYZ*.

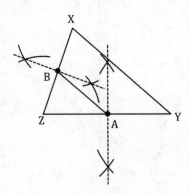

6 The steps are

- Extend $\overline{BA}$.
- Use the "parallel line" construction to construct a line parallel to $\overline{BC}$ and through point A.
- Measure the length BC by making arc 1 with point at B and pencil at C.
- Locate point D by measuring a segment of the same length from point A. D is the fourth vertex of parallelogram $ABCD$.

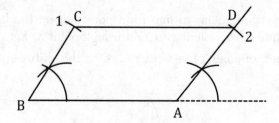

7 Extend $\overline{AB}$, and locate point C so that $AB = BC$ using the "copy a segment" construction. Construct line m, the perpendicular bisector of $\overline{AC}$. Locate point D so that $AB = BD$ using the "copy a segment" construction. $\triangle ABD$ is an isosceles right triangle with angles measuring 45°, 45°, and 90°.

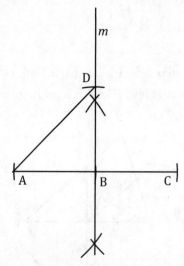

8 Construct an equilateral triangle. Bisect one of its angles to get a pair of 30° angles, and then bisect one of those angles to create two 15° angles.

9 Construct the perpendicular bisectors of $\overline{JK}$ and $\overline{LM}$. The perpendicular bisector of a chord always passes through the center of the circle, so the intersection of the two perpendicular bisectors is point P, the center of the circle.

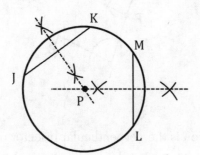

10 The distance from a point to the center is multiplied by the scale factor of the dilation, so $PA' = 2PA$ and $PB' = 2PB$. Construct rays $\overrightarrow{PA}$ and $\overrightarrow{PB}$. With the point at P, measure the distance PA. Move the point to A, and measure to same distance to locate A'. Repeat for point B and B'.

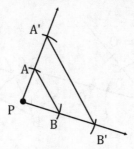

11 Use the "parallel to a line through point off the line" construction to construct a ray through P and parallel to $\overrightarrow{AB}$. With the point at A, measure length AB. Move the point to P, and measure the same distance along the new ray to locate P'. P' is the translation of P by vector $\overrightarrow{AB}$.

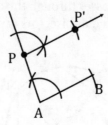

12 The line of reflection is the perpendicular bisector of the segment through a point and its image. Construct the perpendicular bisector of $\overline{PP'}$.

13 Construct a ray through P and Q, and apply the "copy an angle" construction to copy $\angle ABC$ to $\angle PQD$. Then, with the point at P, measure the distance PQ. Use the same distance to locate point Q' along $\overrightarrow{PD}$. Q' is the rotation of Q about point P by $\angle ABC$.

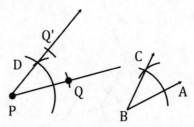

3.4 TRANSFORMATIONS

Transformations can be thought of as functions that operate on geometric figures. The original figure is called the pre-image and the new figure is called the image. There are four basic transformations to be familiar with—translations, reflections, rotations, and dilations. The first three are rigid motions that do not change the size of the figure and result in a congruent figure. The last one, dilations, can change the size of a figure and results in a similar figure.

TRANSLATIONS

A *translation* slides a figure from one position to another. Translations can be specified by a vector, with the notation $T_{\overrightarrow{FG}}$. A vector consists of a magnitude and a direction. The magnitude is represented by the length of the vector and the direction is indicated by following the endpoint, point F, toward the second point. A translation can also be specified by stating one point and the point it maps to. For example, a translation such that maps E to G. The accompanying figure illustrates $T_{\overrightarrow{FG}}(ABCDE) \rightarrow A'B'C'D'E'$. $ABCDE$ is translated a distance FG parallel to the direction of $\overrightarrow{FG}$. The translation is a rigid motion so $ABCDE \cong A'B'C'D'E'$. Also, segments joining corresponding points are congruent and parallel; $\overline{AA'}$ is congruent and parallel to $\overline{BB'}$, $\overline{CC'}$, $\overline{DD'}$, and $\overline{EE'}$.

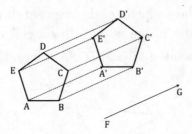

LINE REFLECTIONS

A *line reflection* flips a figure over a line, so that it appears as a mirror image. The line is called the *line of reflection*. The notation r_ℓ indicates a reflection over the line ℓ. The accompanying figure illustrates the reflection of $ABCD$ over line ℓ. Reflections are a rigid motion, so $ABCD \cong A'B'C'D'$. The line of reflection is also the perpendicular bisector of each segment joining corresponding points in the pre-image and image. Therefore, every point on the pre-image and its corresponding point on the image are equidistant from the line of reflection.

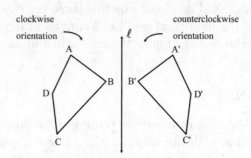

Line reflections do not preserve orientation, which is the direction one must travel to move from one vertex to the next. In the figure, the orientation changes from clockwise to counterclockwise. Noting a change in orientation is one way to identify a reflection.

POINT REFLECTION

A *point reflection* maps each point in the pre-image to a point such that the center of the reflection is the midpoint of the segment through any corresponding pair of points. $\overline{PQ}$ is the image of $\overline{ST}$ after a reflection through point R. Therefore R is the midpoint of $\overline{SP}$ and $\overline{TQ}$. A point reflection is equivalent to a 180° rotation.

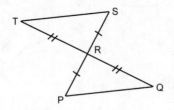

ROTATIONS

A *rotation* is the spinning of a figure about a pivot point called the *center of rotation*. The angle of rotation is measured counterclockwise unless otherwise specified. The accompanying figure shows $ABCD$ rotated about point O to $A'B'C'D'$. The notation $R_{C,a}$ is used to specify a rotation, where a is the angle of rotation and C is the center of rotation. Rotations are rigid motions, so $ABCD \cong A'B'C'D'$. The angle formed by any two corresponding points with the center of rotation as the vertex is equal to the angle of rotation. $m\angle AOA' = m\angle BOB' = m\angle COC' = m\angle DOD' =$ angle of rotation.

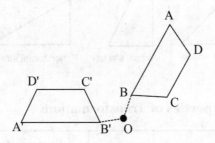

DILATIONS

A *dilation* is a similarity transformation that enlarges or reduces the size of a figure without changing its shape. Angle measures are preserved, and the lengths of segments in the image are proportional to lengths in the pre-image. The ratio of lengths is called the scale factor. Dilations are specified about a center point. The distance from a point to the center is also multiplied by the scale factor after a dilation. $D_{C,k}$ is a dilation with a center point C and scale factor k. The accompanying figure shows the dilation $D_{C,2}(\overline{PQ}) \rightarrow \overline{P'Q'}$.

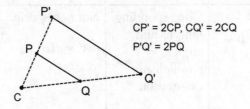

CP' = 2CP, CQ' = 2CQ

P'Q' = 2PQ

HORIZONTAL AND VERTICAL STRETCH

Two examples of transformations that are neither rigid motions nor similarity transformations are the horizontal stretch and vertical stretch. A *horizontal stretch* will elongate a figure only in the horizontal direction. On the coordinate plane, the *x*-coordinate of every point will be multiplied by a scale factor. A *vertical stretch* will do the same thing to the *y*-coordinate.

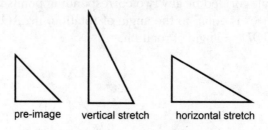

| pre-image | vertical stretch | horizontal stretch |

Summary of the Properties of Transformations

Transformation	What's Preserved	Segments Between Corresponding Points Congruent?	Other Relationships
Reflection $r_{\overleftrightarrow{FE}}(AB) \rightarrow \overline{A'B'}$	Corresponding lengths and angles, parallelism	Not necessarily $AA' \neq BB'$	$\overleftrightarrow{FE}$ is the $\perp$ bisector of $\overline{BB'}$ and $\overline{AA'}$ $\overline{AA'} \parallel \overline{BB'}$
Rotation $R_{C,100°}(AB) \rightarrow \overline{A'B'}$	Corresponding lengths and angles, parallelism, orientation	Not necessarily $AA' \neq BB'$	$\overline{AC} \cong \overline{A'C}$ $\overline{BC} \cong \overline{B'C}$ $\angle ACA' \cong \angle BCB'$

Transformation	What's Preserved	Segments Between Corresponding Points Congruent?	Other Relationships
Translation $T_{\overline{UV}}(\overline{AB}) \rightarrow \overline{A'B'}$	Corresponding lengths and angles, slope, parallelism, orientation	Yes $AA' \neq BB'$	$\overline{AB} \; // \; \overline{A'B'}$ $\overline{AA'} \; // \; \overline{BB'}$
Dilation $D_{C,3/2}(\overline{AB}) \rightarrow \overline{A'B'}$	Corresponding angles, slope, parallelism, orientation	Not necessarily $AA' \neq BB'$	$\overline{AB} \; // \; \overline{A'B'}$

Transformations on the Coordinate Plane

The following set of rules can be used to apply a transformation to a point on the coordinate plane.

Reflection	Rotation About the Origin
$r_{x\text{-axis}}(x, y) \rightarrow (x, -y)$	$R_{90}(x, y) \rightarrow (-y, x)$
$r_{y\text{-axis}}(x, y) \rightarrow (-x, y)$	$R_{180}(x, y) \rightarrow (-x, -y)$
$r_{y=x}(x, y) \rightarrow (y, x)$	$R_{270}(x, y) \rightarrow (y, -x)$
$r_{y=-x}(x, y) = (-y, -x)$	

Translation	Dilation
$T_{h,k}(x, y) \rightarrow (x + h, y + k)$	$D_{k,\text{origin}}(x, y) \rightarrow (kx, ky)$

- Rotations of 180° and 270° can also be found by repeating the rule for a 90° rotation.

COMPOSITIONS

A **composition of transformation** is a sequence of transformations where the image found after the first transformation becomes the pre-image for the next transformation. The symbol ° is used to indicate a composition, as in $T_{3,\,4} \circ D_{3,\,\text{origin}}$. The transformation on the right is performed first. The same composition can also be expressed as $T_{3,\,4}(D_{3,\,\text{origin}})$.

SYMMETRY

A figure has *line symmetry* if a line of reflection can be drawn which divides the figure into two congruent mirror images. For example:

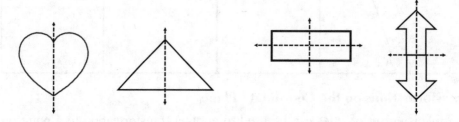

vertical line symmetry   horizontal and vertical line symmetry

A figure has *rotational symmetry* if it can be rotated by some angle $0 < \theta < 360°$ about its center and have every point map to another point on the image. In other words, the image will look identical to the pre-image after the rotation.

The accompanying figure shows rectangle $ABCD$ after rotations of 0°, 90°, 180°, 270°, and 360°. Rotations of 0°, 180°, and 360° result in figures that look identical to the original. We say the figure has 180° rotational symmetry—the 0° and 360° rotations are not considered rotational symmetries. Note that the location of individual points has changed. After a 180° rotation, point A is mapped to point C, B is mapped to D, C is mapped to A, and D is mapped to B.

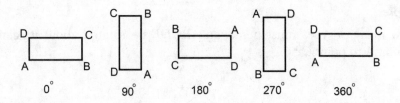

0°   90°   180°   270°   360°

Some figures have more than one rotation that results in an identical figure. An equilateral triangle has 120° and 240° rotational symmetry when rotated about its center.

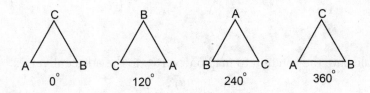

0°   120°   240°   360°

A regular polygon with n sides always has rotational symmetry, with rotations in increments equal to its central angle of $\dfrac{360°}{n}$.

Example

By how many degrees must a regular octagon be rotated so that it maps onto itself? List all the rotations less than 360° that will map an octagon onto itself.

Solution:

$n = 8$ for an octagon.

$$\frac{360°}{n} = \frac{360°}{8} = 45°$$

Any rotation between 0° and 360° that is a multiple of 45° will map the octagon onto itself—45°, 90°, 135°, 180°, 225°, 270°, and 315°.

Example

By how many degrees must pentagon *ABCDE* be rotated about its center to map point *A* to point *C*?

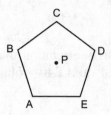

Solution:

First calculate the rotation to map one vertex to the adjacent vertex.

$$\frac{360°}{n} = \frac{360°}{5} = 72°$$

Since rotations are counterclockwise, the pentagon must be rotated three increments of $72°$, or $216°$, to map point *A* to point *C*.

Practice Exercises

1 $\triangle ABC$ undergoes a transformation such that its image is $\triangle A'B'C'$. If the side lengths and angles are preserved, but the slopes of the sides are not, the transformation could be a

(1) reflection or rotation.   (3) translation or rotation.

(2) reflection or translation.   (4) dilation or translation.

2 Which of the following transformations is not a rigid motion?

(1) reflection   (3) rotation

(2) dilation   (4) translation

3 Which of the following letters has more than one line of symmetry?

(1) **A**   (3) **H**

(2) **C**   (4) **L**

4 Triangles AMN and APQ share vertex A. A circle can be constructed that has a center at A and passes through points M, N, P, and Q. If m$\angle MAN =$ m$\angle PAQ$, which of the following is *not* necessarily true?

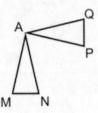

(1) $\angle NAQ$ is a right angle.

(2) $\angle M \cong \angle Q$.

(3) $\triangle APQ$ is the image of $\triangle AMN$ after a rotation about point A.

(4) $\triangle AMN$ and $\triangle AQP$ are isosceles.

5 Which transformation is represented in the figure below?

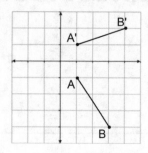

  (1) $R_{\text{origin, }90°}$ (3) $r_{x\text{-axis}}$

  (2) $R_{\text{origin, }180°}$ (4) $r_{y\text{-axis}}$

6 Which of the following will always map a rectangle onto itself?

  (1) reflection over one of its sides

  (2) reflection over one of its diagonals

  (3) rotation of 180° about its center

  (4) rotation of 90° about one of its vertices

7 The area of a rectangle is 4 cm². The rectangle then undergoes a rotation of 90° about one of its vertices, followed by a dilation with a scale factor of 3, centered at the same vertex as the rotation. What is the area of the image?

  (1) 12 cm²   (3) 32 cm²

  (2) 16 cm²   (4) 36 cm²

8 Regular hexagon *ABCDEF* is shown in the figure below. Which of the following transformations will map the hexagon onto itself such that point *D* maps to point *E*?

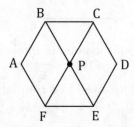

(1) a 60° rotation about point *P* followed by a reflection of line $\overleftrightarrow{BE}$

(2) reflection over line $\overleftrightarrow{BE}$ followed by 60° rotation about point *P*

(3) reflection over line $\overleftrightarrow{CF}$ followed by a translation by vector $\overrightarrow{FE}$

(4) reflection over line $\overleftrightarrow{CF}$ followed by a 120° rotation about point *P*

9 After a certain transformation the image of △*BAD* with coordinates *B*(7, 1) *A*(0, 3) *D*(12, 4) is *B′*(13, 3), *A′*(6, 5), *D′*(18, 1). Is the transformation a translation? Explain why or why not.

10 The vertices of △*MNO* have coordinates *M*(−5, 2), *N*(1, 2), *P*(0, 5). △*M′N′P′* is the image of △*MNP* after a translation. Find the coordinates of *N′* and *P′* if vertex *M* is mapped to *M′*(3, 6).

11 In the figure below, △*DAB* ≅ △*BED* ≅ △*EBC* ≅ △*CFE*. Which of the following sequences of rigid motions will always map △*DAB* to △*CFE*?

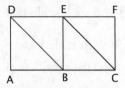

(1) a translation by vector $\overrightarrow{BE}$ followed by a translation by vector $\overrightarrow{EC}$

(2) a rotation of 180° about point *B*, followed by a translation by vector $\overrightarrow{CF}$

(3) a reflection over $\overline{BE}$ followed by a translation by vector $\overrightarrow{BC}$

(4) a translation by vector $\overrightarrow{AC}$ followed by a reflection over $\overline{FC}$

12 Which of the following will map parallelogram *ABCD* onto itself?

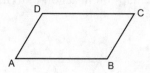

(1) a rotation of 180° about the point of intersection of the two diagonals of *ABCD*

(2) a reflection over $\overline{CD}$, followed by a translation along vector $\overrightarrow{DA}$

(3) a reflection over diagonal $\overline{AC}$

(4) . a reflection over $\overline{BC}$, followed by a translation by vector $\overrightarrow{BA}$

13 In the figure below, △*SDR* ~ △*RCP*. Which of the following compositions of transformations will map △*RCP* to △*SDR*?

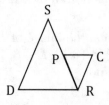

(1) a translation by vector $\overrightarrow{PR}$, followed by a 180° rotation about point *R*, followed by a dilation centered at *R* with scale factor $\dfrac{RS}{RP}$

(2) a dilation centered at *R* with scale factor $\dfrac{RS}{RP}$, followed by a reflection over $\overline{RS}$

(3) a translation by vector $\overrightarrow{CR}$, followed by a reflection over $\overline{PC}$, followed by a dilation centered at *R* with scale factor $\dfrac{RS}{RP}$

(4) a dilation centered at *R* with scale factor $\dfrac{RP}{DS}$, followed by a translation along vector $\overrightarrow{PD}$, followed by a reflection over $\overline{DR}$

14 Graph △*A*(−2, 1), *B*(0, 2), *C*(−1, 4). Graph △*A″B″C″*, the image of △*ABC* after the transformation $T_{1,3} \cdot r_{x=y}$. State the coordinates of *A″B″C″*.

15 Parallelogram *STAR* has coordinates $S(-8, -4)$, $T(-5, -2)$, $A(1, -5)$, and $R(-2, -7)$.

(a) Graph and state the coordinates of $S'T'A'R'$, the image of *STAR* after $r_{y=-1}$.

(b) Graph and state the coordinates of $S''T''A''R''$, the image of $S'T'A'R'$ after $T_{7,-2}$.

(c) Name a single reflection that would map T to T''.

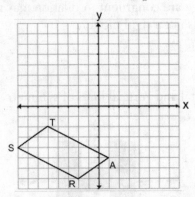

Solutions

1 A translation always preserves slope, so choices (2), (3), and (4) cannot be correct.

The correct choice is **(1)**.

2 Reflections, rotations, and translations are always rigid motions because the image and pre-image are congruent. A dilation may result in an image with a different size, so it is not a rigid motion.

The correct choice is **(2)**.

3 **H** has both horizontal and vertical lines of symmetry.

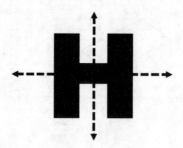

The correct choice is **(3)**.

4 If points M, N, P, and Q all lie on the same circle with center A, then $\overline{AM}$, $\overline{AN}$, $\overline{AP}$, and $\overline{AQ}$ are all radii of circle A, and $AM = AN = AP = AQ$. The triangles are congruent by SAS and are isosceles, so $\angle M \cong \angle N \cong \angle P \cong \angle Q$ and choices (2) and (4) are true.

Since $\angle MAN \cong \angle PAQ$ and $\angle NAP \cong \angle NAP$ by the reflexive property, then $m\angle MAN + m\angle NAP = m\angle NAP + m\angle PAQ$. This is equivalent to $\angle MAP \cong \angle NAQ$. Therefore, P is the rotation of M and Q is the rotation of point N about point A, so one triangle is the rotation of the other. Choice (3) is true. The only choice that cannot be proven is choice (1); we do not know the measures of any of the angles and cannot tell if $\angle NAQ$ measures $90°$.

The correct choice is **(1)**.

5 Approach this problem by simply applying each transformation and checking if $A(1, -1)$ maps to $A'(1, 1)$ and $B(3, -4)$ maps to $B'(4, 3)$. If the correct image results for both, then the choice is correct.

$R_{\text{origin, }90°}(1, -1) \to (1, 1)$ correct	$R_{\text{origin, }90°}(3, -4) \to (4, 3)$
$R_{\text{origin, }180°}(1, -1) \to (-1, 1)$ A' and B' are incorrect	$R_{\text{origin, }180°}(3, -4) \to (-3, 4)$
$R_{x\text{-axis}}(1, -1) \to (1, 1)$ B' is incorrect	$R_{x\text{-axis}}(3, -4) \to (3, 4)$
$r_{y\text{-axis}}(1, -1) \to (-1, -1)$ A' and B' are incorrect	$r_{y\text{-axis}}(3, -4) \to (-3, -4)$

The correct choice is (**1**).

6 Since a rectangle has twofold symmetry, a rotation of $180°$ about its center will map it to itself.

The correct choice is (**3**).

7 Under a dilation, the area of a polygon is proportional to the scale factor squared, so the area becomes 3^2 times larger, or 9 times larger. The new area is $9 \cdot 4 \text{ cm}^2 = 36 \text{ cm}^2$.

The correct choice is (**4**).

8 A reflection over $\overleftrightarrow{BE}$ will map point D to point F. Each central angle measures $\dfrac{360°}{60°} = 60°$. Therefore, a rotation of $60°$ will map any vertex to the adjacent vertex in the counterclockwise direction. The $60°$ rotation will map a point at vertex F to vertex E.

The correct choice is (**2**).

9 If the transformation is a translation, then every point in the pre-image would undergo the same translation $(x, y) \rightarrow (x + h, y + k)$. Calculate each h and k by working backwards from the pre-image and image:

$B'(13, 3) = B(7 + h, 1 + k)$	$13 = 7 + h, h = 6$	$3 = 1 + k, k = 2$
$A'(6, 5) = A(0 + h, 3 + k)$	$6 = 0 + h, h = 6$	$5 = 3 + k, k = 2$
$D'(18, 1) = D(12 + h, 4 + k)$	$18 = 12 + h, h = 6$	$1 = 4 + k, k = -3$

The transformation is not a translation because A and B undergo the translation $T_{6,2}$ while D undergoes the translation $T_{6,-3}$.

10 First, find the translation rule.

$T_{h,k} \, M(-5, 2) \rightarrow M'(-5 + h, 2 + k)$

Using the coordinates $(3, 6)$ for M', solve for h and k.

$-5 + h = 3$	$2 + k = 6$
$h = 8$	$k = 4$

The translation is $T_{8,4}$. Now apply that translation to N and P.

$T_{8,4} \, N(1, 2) \rightarrow N'(1 + 8, 2 + 4) \rightarrow N'(9, 6)$
$T_{8,4} \, P(0, 5) \rightarrow P'(0 + 8, 5 + 4) \rightarrow P'(8, 9)$

The correct choice is **(3)**.

11 The figure shows the $\triangle DAB$ after a $180°$ rotation about B. A translation by $\overrightarrow{CF}$ will then slide the image up and map it to $\triangle CFE$.

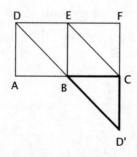

The correct choice is **(2)**.

12 Rotating 180° about the point of intersection of the diagonals will map:
$A \rightarrow C, B \rightarrow D, C \rightarrow A$, and $D \rightarrow B$.

The other choices will not map $ABCD$ onto itself. Choices (2), (3), and (4) result in the following:

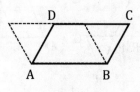

choice (2)

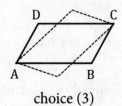

choice (3)

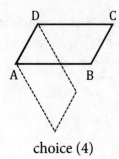

choice (4)

The correct choice is **(1)**.

13 The figures show the figure after the translation by $\overrightarrow{PR}$, and then the rotation of $180°$ about R. The final dilation will map C'' to D and R'' to S.

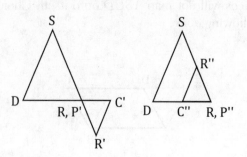

The correct choice is (**1**).

14 In a composition always do the rightmost transformation first. For $T_{1,3} \cdot r_{x=y}$, start with $r_{y=x}$. For a reflection over the line $y = x$, switch the order of the coordinates.

$r_{y=x}$: $A(-2, 1) \rightarrow A'(1, -2)$ $B(0, 2) \rightarrow B'(2, 0)$ $C(-1, 4) \rightarrow C'(4, -1)$

Now use the image from the reflection as the pre-image of the translation.

For a translation $T_{1,3}$, add 1 to each x-coordinate and 3 to each y-coordinate.

$T_{1,3}$: $A'(1, -2) \rightarrow A''(2, 1)$ $B'(2, 0) \rightarrow B''(3, 3)$ $C'(4, -1) \rightarrow C''(5, 2)$

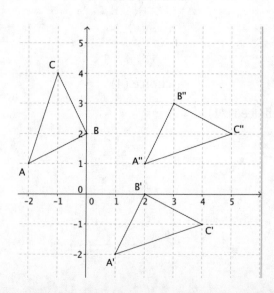

15 a) The x-coordinates of each point in *STAR* are unchanged after a reflection over the line $y = -1$. To find the y-coordinates, count the vertical distance from the point to $y = -1$, then continue by the same distance on the other side of $y = -1$. $S(-8, -4)$ is 3 units below $y = -1$.

The y-coordinate of S' is $-1 + 3 = 2$. $T(-5, -2)$ is 1 unit below $y = -1$.	$S'(-8, 2)$
The y-coordinate of T' is $-1 + 1 = 0$. $A(1, -5)$ is 4 units below $y = -1$.	$T'(-5, 0)$
The y-coordinate of A' is $-1 + 4 = 3$. $R(-2, -7)$ is 6 units below $y = -1$.	$A'(1, 3)$
The y-coordinate of R' is $-1 + 6 = 5$. $S'(-8, 2)\ T'(-5, 0)\ A'(1, 3)\ R'(-2, 5)$	$R'(-2, 5)$

b) Apply the translation $T_{7,-2}$ by adding 7 to each x-coordinate and subtracting 2 from each y-coordinate.

$$S''(-1,0)\,T''(2,-2)\,A''(8,1)\,R''(5,3)$$

c) To reflect $T(-5, -2)$ to $T''(2, -2)$, note that the y-coordinates are the same, so we need a reflection over a vertical line that is halfway between $x = -5$ and $x = 2$.

$$\frac{1}{2}(-5 + 2) = -1\frac{1}{2}$$

A reflection over the line $x = -1\frac{1}{2}$ will map $T(-5, -2)$ to $T''(2, -2)$.

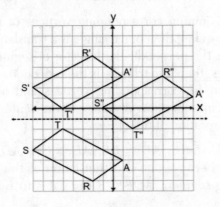

3.5 TRIANGLE CONGRUENCE

CONGRUENCE

Two figures are congruent if one figure can be mapped onto another by a sequence of rigid motions. The congruent figures will have exactly the same size and shape, which means for polygons that all *corresponding* pairs of sides and angles are congruent.

A congruence statement will match up the corresponding pairs of congruent parts.

For example, from the statement $\triangle BUS \cong \triangle CAR$ we can conclude:

$$\angle B \cong \angle C \quad \overline{BU} \cong \overline{CA}$$
$$\angle U \cong \angle A \quad \overline{US} \cong \overline{AR}$$
$$\angle S \cong \angle R \quad \overline{SB} \cong \overline{RC}$$

Once two figures have been proven congruent, we know all pairs of parts are congruent. In triangles, this theorem is abbreviated by CPCTC—corresponding parts of congruent triangles are congruent.

PROVING POLYGONS CONGRUENT BY TRANSFORMATIONS

Polygons can be proven congruent by identifying a sequence of rigid motions that map one onto the other. One method is to show each vertex of the pre-image maps to a corresponding vertex of the image using the same transformation, or sequence of transformations. The angle and segment relationships between pre-image, image, lines of reflection, center of rotation, and translation vectors can help establish specific transformations that map one point onto another:

- If the segments joining corresponding vertices of the two triangles are congruent and parallel, then one triangle is a translation of the other.
- If a single line is the perpendicular bisector of segments formed by corresponding vertices, then one triangle is the reflection of the other.
- If angles formed by corresponding vertices and a center point are all congruent, and corresponding distances to the center point are equal, then one triangle is a rotation of the other.

For example, given $\overline{ADBE}$, $\overline{AD} \cong \overline{BE}$, $\overline{AD} \cong \overline{CF}$, $\overline{AD} // \overline{CF}$, we can show that $\triangle ABC \cong \triangle DEF$ because one is a translation of the other. Point A translates to point D along segment $\overline{ADBE}$. B translates to E by the same distance along the same segment, and C translates by the same distance along a parallel segment. Each point in $\triangle ABC$ is mapped to a corresponding point in $\triangle DEF$ by the same translation (distance AD along a vector parallel to $\overline{ADBE}$). Therefore, $\triangle DEF$ is a translation of $\triangle ABC$, and the two triangles are congruent because translations are a rigid motion.

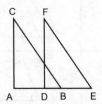

TRIANGLE CONGRUENCE POSTULATES

We know two polygons are congruent if all pairs of corresponding sides and angles are congruent. When proving two triangles congruent we do not need to prove all three pairs of sides and all three pairs of angles congruent. There are five shortcuts commonly used—the SAS, SSS, ASA, AAS, and the HL criteria. Each criteria requires only a total of three pairs of parts to be congruent for us to conclude the triangles are congruent.

Criteria	What It Looks Like
SAS—two pairs of sides and the included angles are congruent.	
SSS—three pairs of sides are congruent.	
ASA—two pairs of angles and the included sides are congruent.	
AAS—two pairs of angles and the nonincluded sides are congruent.	
HL—the hypotenuses and one pair of legs are congruent in a pair of right triangles.	

Once a pair of triangles has been proven congruent using one of these postulates, then CPCTC can be used to justify why any of the remaining pairs of corresponding parts are congruent.

You do not need to limit yourself strictly to a rigid motion or a congruence postulate. Sometimes the given information will lead to a combination of these approaches.

STRATEGIES FOR WRITING CONGRUENCE PROOFS

1. Mark congruent parts on the figure using given information and anything else you can conclude from the figure.
2. Look for the following common relationships:
   a. Vertical angles
   b. Shared sides and angles
   c. Linear pairs
   d. Supplementary and complementary angles
   e. Parallel lines and perpendicular lines
   f. Midpoints and bisectors of segments
   g. Angle bisectors
   h. Altitudes, medians, and perpendicular bisectors in triangles
   i. Isosceles triangle theorem, exterior angle theorem, and triangle angle sum theorem
3. Statements should always refer to specific named parts of the figure (such as $\triangle ABC$, $\overline{AB}$, $\angle C$). Reasons may only involve definitions, postulates, and theorems. Never name a specific part of the figure in the reasons column.
4. There must be a line in your proof for each part of the postulate you use. You can label each line that corresponds to a part of the postulate with (A) or (S), or use checkboxes:

S	A	S
✓	✓	✓

OVERLAPPING TRIANGLES

Angle addition or segment addition can be used to show parts are congruent when triangles overlap. For example, if given $\overline{ABCD}$ and $\overline{AB} \cong \overline{CD}$, we can prove $\overline{AC} \cong \overline{BD}$ by stating that $\overline{BC}$ is congruent to itself by the reflexive property, then adding the two congruence statements:

$$\overline{BC} \cong \overline{BC} \qquad \text{reflexive property}$$

$$\overline{AC} \cong \overline{BD} \qquad \text{given}$$

$$\overline{AB} + \overline{BC} \cong \overline{BC} + \overline{CD} \qquad \text{addition property}$$

$$\overline{AC} \cong \overline{BD} \qquad \text{partition property}$$

A   B   C   D

Another strategy is to sketch the two triangles separately. By pulling them apart, you may more easily see the parts required to complete the proof.

Practice Exercises

1 $A'B'C'D'$ is the image of $ABCD$ after a rotation about the origin, and $A''B''C''D''$ is the image of $A'B'C'D'$ after a translation. Which of the following must be true?

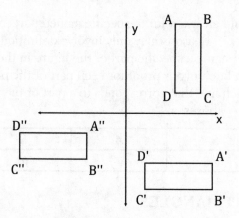

(1) All three figures are rectangles.

(2) $ABCD$ is congruent to $A'B'C'D'$, but not congruent to $A''B''C''D''$.

(3) $ABCD$ is congruent to $A''B''C''D''$, but not congruent to $A'B'C'D'$.

(4) $ABCD$ is congruent to both $A'B'C'D'$ and $A''B''C''D''$.

2 Which triangle congruence criteria can be used to prove the two triangles are congruent?

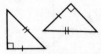

(1) SAS

(3) SSS

(2) ASA

(4) HL

3 Which of the following pieces of information would *not* allow you to conclude that $\triangle ACD \cong \triangle ACB$?

(1) $\overline{AC}$ bisects $\angle BCD$ and $\overline{CD} \cong \overline{CB}$.

(2) $\overline{AC}$ bisects $\angle BCD$ and $\overline{AD} \cong \overline{AB}$.

(3) $\overline{AC}$ bisects $\angle BCD$ and $\angle BAD$.

(4) $\overline{AC}$ bisects $\angle BAD$ and $\overline{AD} \cong \overline{AB}$.

4 Which of the following is sufficient to prove △*SMA* ≅ △*S'M'A'*?

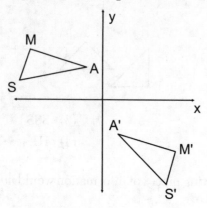

(1) There is a sequence of rigid motions that maps
       ∠*M* to ∠*M'* and ∠*S* to ∠*S'*.

(2) There is a sequence of rigid motions that maps
       ∠*M* to ∠*M'* and $\overline{MS}$ to $\overline{M'S'}$.

(3) There is a sequence of rigid motions that maps
       ∠*S* to ∠*S'* and $\overline{MS}$ to $\overline{M'S'}$.

(4) There is a sequence of rigid motions that maps
       ∠*A* to ∠*A'* and point *A* to point *A'*.

5 Given: $\overline{AB}$ // $\overline{CD}$, $\overline{AB}$ ≅ $\overline{CD}$

   Prove: △*ABE* ≅ △*CDE*

6 Given: $\overline{BC}$ bisects $\overline{AD}$ at *E*, ∠*A* ≅ ∠*D*

   Prove: △*ABE* ≅ △*DCE*

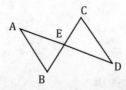

7 *ABCDEF* is a regular hexagon. State a sequence of rigid motions that could be used to justify why △*ABC* is congruent to △*AEF*.

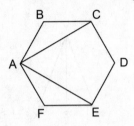

8 In the accompanying figure of △*FGH* and △*GIJ*, there is a sequence of rigid motions that maps *F* to *G*, *G* to *I*, and *H* to *J*. Explain why ∠*H* must be congruent to ∠*J*.

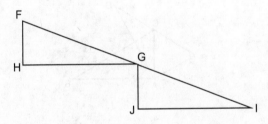

9 Given $\overline{RPB}$ and line *m* is the perpendicular bisector of $\overline{TJ}$ and $\overline{RB}$, explain why △*TRP* ≅ △*JBP* and why $\overline{TR}$ ≅ $\overline{JB}$.

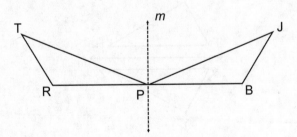

10 Given: $\triangle WXY$ and perpendicular bisector $\overline{YZ}$

Explain why $\triangle WZY \cong \triangle XZY$ in terms of a rigid motion.

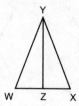

11 Given: $\overline{AB} \cong \overline{AD}$, $\overline{AC} \cong \overline{AE}$, m$\angle BAD = 60°$, and m$\angle CAE = 60°$

Prove: $\triangle BAC \cong \triangle DAE$ in terms of a rigid motion

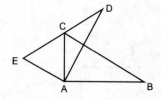

12 Given: line m is the perpendicular bisector of $\overline{FX}$, $\overline{GY}$, and $\overline{HZ}$

Prove: $\triangle XYZ \cong \triangle FHG$

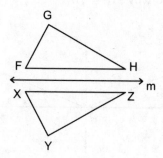

Solutions

1 Both rotations and translations are rigid motions, and any sequence of rigid motions will result in a congruent image. Therefore, $ABCD$ is congruent to both $A'B'C'D'$ and $A''B''C''D''$.

The correct answer choice is (**4**).

2 Both triangles are right triangles, a corresponding pair of legs are congruent, and the two hypotenuses are congruent. Therefore HL applies. SAS is not correct here because the angle does not lie between the two congruent sides.

The correct choice is (**4**).

3 The shared side $\overline{AC}$ is a pair of congruent sides. Check each choice to find one that does *not* provide enough information to complete one of the congruent postulates.

Choice 1: The angle bisector gives $\angle DCA \cong \angle BCA$, and SAS applies.

Choice 2: The angle bisector gives $\angle DCA \cong \angle BCA$, and we have two sides and an angle, but the angle is not the included angle. None of the congruence postulates apply.

Choice 3: The angle bisector gives $\angle DCA \cong \angle BCA$ and $\angle DAC \cong \angle BAC$. ASA applies.

Choice 4: The angle bisector gives $\angle DAC \cong \angle BAC$, and SAS applies.

The correct choice is (**2**).

4 Choice 1 maps two consecutive angles onto corresponding angles, as well as M to M' and S to S'. This means $\overline{SM} \cong \overline{S'M'}$ because a sequence of rigid motions maps the endpoints of $\overline{SM}$ onto corresponding endpoints. $\overline{SM}$ is the included side between the congruent angles, therefore, $\triangle SMA \cong \triangle S'M'A'$ by the SAS postulate.

The correct choice is (**1**).

5

Statement	Reason
1. $\overline{AB} \cong \overline{CD}$, $\overline{AB} \,/\!/\, \overline{CD}$	1. Given
2. $\angle A \cong \angle C$ $\angle B \cong \angle D$	2. Alternate interior angles formed by the parallel lines are congruent
3. $\triangle ABE \cong \triangle CDE$	3. ASA

6

Statement	Reason
1. $\overline{BC}$ bisects $\overline{AD}$ at E	1. Given
2. E is the midpoint of $\overline{AD}$	2. A bisector intersects a segment at its midpoint
3. $\overline{AE} \cong \overline{ED}$	3. A midpoint divides a segment into two congruent segments
4. $\angle AEB \cong \angle DEC$	4. Vertical angles are congruent
5. $\angle A \cong \angle D$	5. Given
6. $\triangle ABE \cong \triangle DCE$	6. ASA

7 $\triangle ABC$ is mapped to $\triangle AFE$ by a rotation about point A followed by a reflection over $\overline{AE}$. The angle of rotation is $\angle CAE$. From the figure each of the diagonals $\overline{AC}$, $\overline{AE}$, and $\overline{CE}$ are congruent, so $\triangle ACE$ is equilateral, and its angles measure 60°. The angle of rotation is 60°. The sequence of rigid motions is a 60° rotation about point A followed by a reflection over $\overline{AE}$.

8 The two triangles are congruent because there exists a rigid motion that maps each vertex of $\triangle JGH$ to a corresponding vertex $\triangle GIJ$. Since the triangles are congruent, all pairs of corresponding parts are congruent (CPCTC). $\angle G$ and $\angle H$ are congruent corresponding angles.

9 Since line m is the perpendicular bisector of both $\overline{TJ}$ and $\overline{RB}$, it is a line of reflection that maps T to J and R to B. The same reflection will map P to itself because it lies on the line of reflection. A single reflection maps $\triangle TRP$ onto $\triangle JBP$, and reflections are rigid motions, so the two triangles must be congruent. $\overline{TR} \cong \overline{JB}$ because corresponding parts of congruent triangles are congruent.

10 Point Y is the image of itself after a reflection over $\overline{YZ}$ because Y lies on the line of reflection. Point Z is the image of itself for the same reason. X is the image of point W after a reflection over $\overline{YZ}$ because a line of reflection is the perpendicular bisector of the segment joining the pre-image and the image after the reflection. Therefore, the triangles are congruent because one is mapped to the other by a reflection over $\overline{YZ}$, which is a rigid motion.

11 $\overline{AB} \cong \overline{AD}$; therefore, D is the image of B after a 60° rotation about A. $\overline{AC} \cong \overline{AE}$; therefore, C is the image of E after a 60° rotation about A. Point A will map to itself after the same rotation because it is the center of rotation. Therefore, $\triangle BAC$ is the image of $\triangle DAE$ after a 60° rotation about A. $\triangle BAC \cong \triangle DAE$ because rotations are rigid motions.

12 A perpendicular bisector of a segment is also the line of reflection that maps one endpoint to the other. Therefore, after a reflection over line m, X is the image of F, Z is the image of H, and Y is the image of G. $\triangle FHG$ is mapped to $\triangle XZY$ by a reflection, which is a rigid motion; therefore, $\triangle XZY \cong \triangle FHG$.

3.6 COORDINATE GEOMETRY

COORDINATE GEOMETRY FORMULAS

Given two points (x_1, y_1) and (x_2, y_2):

- The distance between the points is $\sqrt{(x_2 - x_1)^2 + (y_2 - y_1)^2}$

- The midpoint of the segment joining the points is $\left(\dfrac{x_1 + x_2}{2}, \dfrac{y_1 + y_2}{2} \right)$

- The slope of the segment joining the points is $\dfrac{y_2 - y_1}{x_2 - x_1}$, or $\dfrac{rise}{run}$

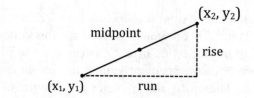

DIVIDING A SEGMENT PROPORTIONALLY

A directed segment is one that has a specified starting point and ending point. We can divide a directed segment into parts that are in any given ratio using the two proportions ratio $= \dfrac{x - x_1}{x_2 - x}$ and ratio $= \dfrac{y - y_1}{y_2 - y}$.

For example, given $J(1, -2)$ and $K(11, 3)$, find the coordinate (x, y) of point L that divides $\overline{JK}$ in a 2:3 ratio.

J(1, -2)   L(x,y)   K(11, 3)

- Use the coordinates of J for point 1 and the coordinates of K for point 2.

$$x_1 = 1 \qquad x_2 = 11 \qquad y_1 = -2 \text{ and } y_2 = 3$$

- Apply the formula for the x-coordinate.

$$\text{ratio} = \frac{x - x_1}{x_2 - x}$$

$$\frac{2}{3} = \frac{x - 1}{11 - x}$$
$$2(11 - x) = 3(x - 1)$$
$$22 - 2x = 3x - 3$$
$$25 = 5x$$
$$x = 5$$

- Repeat the process for the y-coordinate.

$$\text{ratio} = \frac{y - y_1}{y_2 - y}$$

$$\frac{2}{3} = \frac{y - (-2)}{3 - y}$$
$$3y + 6 = 6 - 2y$$
$$5y = 0$$
$$y = 0$$

Point $L(5, 0)$ divides $\overline{JK}$ in a 2:3 ratio.

AREA AND PERIMETER

The lengths found with the distance formula can be used to calculate the perimeter and area of figures. If the figure is irregular, three strategies can be used:

- Divide the figure into shapes whose areas can be calculated easily (squares, rectangles, triangles, trapezoids, and circles).
- Sketch a bounding rectangle around the figure. Calculate the area of the rectangle, and then subtract the area of all the triangles that fall outside the figure but with the rectangle.
- If the figure has curves, estimate the area by modeling the curved sides with straight segments. The more segments that are used to model a curve, the more accurate the result will be.

Example

Find the area of polygon $ABCDE$, with vertices $A(-3, -5)$, $B(2, -5)$, $C(5, 3)$, $D(-2, 5)$, and $E(-5, -1)$.

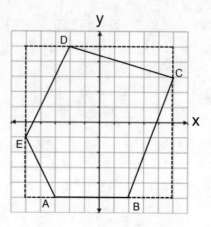

Solution:

Sketch the bounding rectangle in around $ABCDE$.

The length is 10 and the width is 10, giving an area of 100. The triangles have areas:

$$\text{upper left triangle} = \frac{1}{2}(3 \cdot 6) = 9$$

$$\text{upper right triangle} = \frac{1}{2}(7 \cdot 2) = 7$$

$$\text{lower left triangle} = \frac{1}{2}(4 \cdot 2) = 4$$

$$\text{lower right triangle} = \frac{1}{2}(3 \cdot 8) = 12$$

The area of $ABCDE = 100 - 9 - 7 - 4 - 12 = 68$

COLLINEARITY

Three points are **collinear** if the slopes between any two pairs are equal. For example, points A, B, and C are collinear if the slope of $\overline{AB}$ equals the slope of $\overline{BC}$.

EQUATIONS OF LINES

- The slopes of parallel lines are equal.
- The slopes of perpendicular lines are negative reciprocals. (If the slope of line m is $\frac{2}{3}$, then the slope of any line perpendicular to m is $-\frac{3}{2}$.)
- The equation of a line in slope-intercept form is $y = mx + b$ where m is the slope and b is the y-intercept. To graph the line, plot a point on the y-axis at the y-intercept. From that point, plot additional points using the rise and run from the slope.
- The equation of a line in point-slope form is $y - y_1 = m(x - x_1)$ where m is the slope and (x_1, y_1) are the coordinates of any point on the line. To graph the line, plot the first point at (x_1, y_1). From that point, plot additional points using the rise and run from the slope.

Strategy for writing the equation of a line in slope-intercept form:

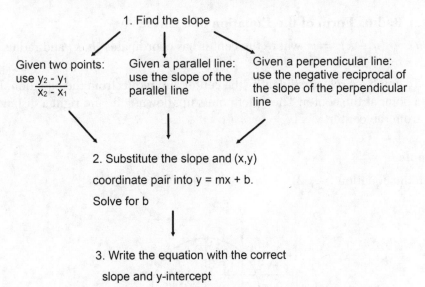

TRANSFORMATIONS AND LINES

Translations and Dilations

Translations and dilations preserve slope, so the slope of the image will be the same as the slope of the pre-image.

To translate or dilate a line given its equation,

1. Choose any point on the line (the y-intercept is often an easy choice).

2. Apply the translation or dilation to that point.

3. Find the equation of the line that has the same slope as the original line and passes through the transformed point.

Rotations

Rotations of 90° will result in a line perpendicular to the original, so the slope will be the negative reciprocal. To write the equation of a line after a 90° rotation, use the same procedure for translations and dilations, except use the negative reciprocal of the slope.

EQUATION OF THE CIRCLE

Center Radius Form of the Equation of a Circle

$(x - h)^2 + (y - k)^2 = r^2$ where the center has coordinates (h, k) and radius has length r.

- To graph a circle, first identify the center and radius from the equation. Plot a point at the center. Then plot points up, down, left, and right a distance r from the center.

Example

Graph the equation $(x - 2)^2 + (y + 1)^2 = 9$.

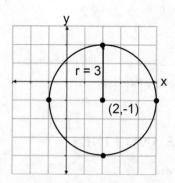

The center is located at $(2, -1)$, and $r^2 = 9$, so $r = 3$. We plot the center point at $(2, -1)$; then plot points up, down, right, and left 3 units from the center. Use these four points as a guide to complete the circle.

General Form of the Equation of a Circle

$$x^2 + y^2 + Cx + Dy + E = 0$$

To find the coordinates of the center and the radius from the general form of the equation, you will need to convert it to the center-radius form using the following procedure:

1. Group the x-terms and y-terms on one side of the equation, and the constant on the other side of the equation.

2. Complete the square with the x-terms, and then complete the square with the y-terms.

Example

Find the coordinates of the center and the length of the radius of a circle whose equation is $x^2 + 4x + y^2 - 6y + 7 = 0$.

Solution:

Bring the constant term to the right.

$$x^2 + 4x + y^2 - 6y = -7$$

The coefficient of x is 4, so a constant term of $\left(\dfrac{4}{2}\right)^2$, or 4, is needed to complete the square with the x-terms. The coefficient of y is -6, so a constant term of $\left(\dfrac{-6}{2}\right)^2$, or 9, is needed to complete the square with the y-terms.

$$x^2 + 4x + 4 + y^2 - 6y + 9 = -7 + 4 + 9$$

$$(x + 2)^2 + (y - 3)^2 = 6$$

The center has coordinates $(-2, 3)$ and the radius has a length of $\sqrt{6}$.

Practice Exercises

1 Points $A(2, -1)$ and $B(8, -3)$ lie on line m. After a rotation of $90°$ about the origin, the images of A and B are A' and B'. If A' and B' lie on line n, what is the equation of line n?

(1) $y = -3x + 2$

(3) $y = 3x - 1$

(2) $y = -\dfrac{1}{3}x - \dfrac{1}{3}$

(4) $y = \dfrac{1}{3}x + 6$

2 What is the equation of the line $6x + 2y = 12$ after a dilation by a scale factor of 5?

(1) $y = -3x + 30$

(3) $y = -15x + 30$

(2) $y = -3x + 6$

(4) $y = -15x + 6$

3 Which of the following lines is perpendicular to the line $x + 4y = 8$?

(1) $y = -\dfrac{1}{4}x + 2$

(3) $y = \dfrac{1}{4}x + 2$

(2) $y = 4x + 3$

(4) $y = -4x + 3$

4 Which of the following is the equation of a line parallel to $2x + 3y + 6 = 0$ and passes through the point $(6, 1)$?

(1) $y = -\dfrac{2}{3}x + 5$

(3) $y = -2x + 13$

(2) $y = 2x - 11$

(4) $y = -\dfrac{2}{3}x + 3$

5 What are the coordinates of the midpoint of a segment whose endpoints have coordinates $(3, 1)$ and $(15, -7)$?

(1) $(27, -15)$ (3) $(6, -4)$

(2) $(-6, 4)$ (4) $(9, -3)$

6 The diameter of a circle has endpoints with coordinates $(4, -1)$ and $(8, 3)$. Which of the following is an equation of the circle?

(1) $(x - 2)^2 + (y - 1)^2 = 8$ (3) $(x - 6)^2 + (y - 1)^2 = 8$

(2) $(x - 2)^2 + (y - 1)^2 = 32$ (4) $(x - 6)^2 + (y - 1)^2 = 32$

7 Are the segments $\overline{AB}$ and $\overline{TU}$ congruent, given coordinates $A(1, 4), B(-3, 6)$, $T(2, 5)$, and $U(4, 1)$? Justify your answer.

8 Find the coordinates of the point W that divides directed segment $\overline{UV}$ in a 1:5 ratio, given coordinates $U(-3, 7)$ and $V(9, 1)$.

9 Point A has coordinates $(-2, 7)$ and point B has coordinates $(6, 3)$. Line m has the property that every point on the line is equidistant from points A and B. Find the equation of line m.

10 A circle is described by the equation $x^2 + 6x + y^2 - 12y + 25 = 0$. Find the radius of the circle and the coordinates of its center.

11 Circle P has center $P(4, -5)$ and a radius with length $\sqrt{65}$. Does the point $A(8, 2)$ lie on circle P? Justify your answer.

12 Parallelogram $ABCD$ has coordinates $A(2, -1), B(5, 1), C(a, b)$, and $D(3, 4)$. Write the equation of the line that contains side $\overline{CD}$.

13 Estimate the area of the ellipse shown in the accompanying figure by modeling it as the sum of 5 rectangles.

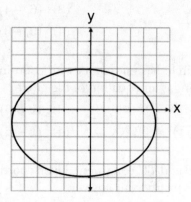

14 Ben wants to estimate the area of the curved region shown. He would like to do so by calculating the average of the areas of two different circles.

a) Graph the two circles that Ben could use.

b) Using the two circles from part (a), estimate the area of the curved region. Each grid unit represents 1 centimeter. Round to the nearest 1 cm².

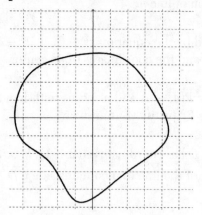

15 $\triangle SKY$ has coordinates $S(0, 2)$, $K(6, 0)$, $Y(4, 4)$.

a) Find the equation of the perpendicular bisector to side $\overline{SK}$.

b) Show that the perpendicular bisector is also an altitude.

Solutions

1 Applying a 90° rotation, A maps to $A'(1, 2)$ and B maps to $B'(3, 8)$.

$$\text{Slope } \overleftrightarrow{A'B'} = \frac{y_2 - y_1}{x_2 - x_1}$$

$$= \frac{8 - 2}{3 - 1}$$

$$= 3$$

Write the equation of the line using the point-slope form and the coordinates of A' for x_1 and y_1.

$$y - y_1 = m(x - x_1)$$
$$y - 2 = 3(x - 1)$$
$$y - 2 = 3x - 3$$
$$y = 3x - 1$$

The correct choice is **(3)**.

2 A dilation with a positive scale factor will preserve slope, but the distance of each point from the center will be multiplied by the scale factor. The strategy is to rewrite the equation in $y = mx + b$ form. The dilated line will have the same slope, but the y-intercept will be multiplied by the scale factor.

$$6x + 2y = 12$$
$$2y = -6x + 12$$
$$y = 3x + 6$$

The y-intercept of the original line is 6, and the y-intercept of the dilated line is $6 \cdot 5$, or 30.

$$y = -3x + 30$$

The correct choice is **(1)**.

3 Perpendicular lines have negative reciprocal slopes. Rewrite the equation in slope-intercept form to identify the slope.

$$x + 4y = 8$$
$$4y = -x + 8$$
$$y = -\frac{1}{4}x + 2$$

The negative reciprocal of $-\frac{1}{4}$ is 4, so the perpendicular line must have a slope of 4. The y-intercept can be any value since a particular perpendicular line was not specified. The only line with a slope of 4 is $y = 4x + 3$.

The correct choice is **(2)**.

4 Parallel lines have equal slopes, so the first step is to rewrite the equation in slope-intercept form to help identify the slope.

$$2x + 3y + 6 = 0$$
$$3y = -2x - 6$$
$$y = -\frac{2}{3}x - 2$$

The slope is $-\frac{2}{3}$.

Now substitute $x = 6$, $y = 1$, and the slope into $y = mx + b$ and solve for the new y-intercept.

$$y = mx + b$$
$$y = -\frac{2}{3}x + b$$
$$1 = -\frac{2}{3}(6) + b$$
$$1 = -4 + b$$
$$b = 5$$

The equation is $y = -\frac{2}{3}x + 5$.

The correct choice is **(1)**.

5 Apply the midpoint formula:

$$x_{MP} = \frac{1}{2}(x_1 + x_2) \quad y_{MP} = \frac{1}{2}(y_1 + y_2)$$

$$x_{MP} = \frac{1}{2}(3 + 15) \quad y_{MP} = \frac{1}{2}(1 + (-7))$$

$$x_{MP} = \frac{1}{2}(18) \quad y_{MP} = \frac{1}{2}(-6)$$

$$x_{MP} = 9 \quad y_{MP} = -3$$

The correct choice is **(4)**.

6 The center and radius of the circle are needed to write its formula. The midpoint of the diameter gives the center:

$$x_{MP} = \frac{1}{2}(x_1 + x_2) \quad y_{MP} = \frac{1}{2}(y_1 + y_2)$$

$$x_{MP} = \frac{1}{2}(4 + 8) \quad y_{MP} = \frac{1}{2}(-1 + 3)$$

$$x_{MP} = \frac{1}{2}(12) \quad y_{MP} = \frac{1}{2}(2)$$

$$x_{MP} = 6 \quad y_{MP} = 1$$

The radius is the distance from the center point to either endpoint of the diameter. Apply the distance formula with points $(6, 1)$ and $(8, 3)$.

$$\text{distance} = \sqrt{(x_1 - x_2)^2 + (y_1 - y_2)^2}$$

$$= \sqrt{(8 - 6)^2 + (3 - 1)^2}$$

$$= \sqrt{(2)^2 + (2)^2}$$

$$= \sqrt{8}$$

The radius of the circle is $\sqrt{8}$, and its center has coordinates $(6, 1)$. Substitute these values for $r, h,$ and k in the equation of a circle:

$$(x - h)^2 + (y - k)^2 = R^2$$
$$(x - 6)^2 + (y - 1)^2 = \sqrt{8}^2$$
$$(x - 6)^2 + (y - 1)^2 = 8$$

The correct choice is **(3)**.

7 Two segments are congruent if their lengths are equal, so apply the distance formula to determine the length of each segment.

$$d = \sqrt{(x_1 - x_2)^2 + (y_1 - y_2)^2}$$

$$AB = \sqrt{(-3 - 1)^2 + (6 - 4)^2} \qquad TU = \sqrt{(4 - 2)^2 + (1 - 5)^2}$$
$$= \sqrt{(-4)^2 + (2)^2} \qquad\qquad = \sqrt{(2)^2 + (-4)^2}$$
$$= \sqrt{16 + 4} \qquad\qquad\qquad = \sqrt{4 + 16}$$
$$= \sqrt{20} \qquad\qquad\qquad\quad = \sqrt{20}$$

$AB = TU$; therefore, the 2 segments are congruent.

8

$$\frac{UW}{WV} = \frac{1}{5} = \frac{x - (-3)}{9 - x}$$
$$9 - x = 5(x + 3)$$
$$9 - x = 5x + 15$$
$$-6 = 6x$$
$$x = -1$$

Repeating for the y-coordinate:

$$\frac{UW}{WV} = \frac{1}{5} = \frac{y - 7}{1 - y}$$
$$1 - y = 5(y - 7)$$

The coordinates of W are $(-1, 6)$.

9 Line m is the line of reflection that maps A to B, so it must be the perpendicular bisector of $\overline{AB}$. To find the perpendicular bisector, calculate the midpoint and slope of $\overline{AB}$. Then write the equation of the line with the negative reciprocal slope that passes through the midpoint.

$$x_{MP} = \frac{1}{2}(x_1 + x_2), \quad y_{MP} = \frac{1}{2}(y_1 + y_2)$$

$$x_{MP} = \frac{1}{2}(-2 + 6), \quad y_{MP} = \frac{1}{2}(7 + 3)$$

$$x_{MP} = 2, \quad y_{MP} = 5$$

$$\text{slope} = \frac{y_2 - y_1}{x_2 - x_1}$$

$$\text{slope}_{AB} = \frac{3 - 7}{6 - (-2)}$$

$$= -\frac{1}{2}$$

$\text{slope}_{\text{line } m} = 2$ $\perp$ lines have negative reciprocal slopes

$$y - y_1 = m(x - x_1)$$ point-slope equation of a line

substitute the coordinates

$$y - 5 = 2(x - 2) \text{ or } y = 2x + 1$$ of the midpoint for x_1 and y_1, and 2 for m

10 Apply the completing the square procedure to rewrite the circle in $(x - h)^2 + (y - k)^2 = R^2$ form. Rewrite the equation with the variables on the left and constant on the right.

$$x^2 + 6x + y^2 - 12y + 25 = 0$$
$$x^2 + 6x + y^2 - 12y = -25$$

The constant needed to complete the square is $\left(\frac{1}{2}b\right)^2$, where b is the coefficient of the linear x- and y-terms. For the x-terms, the necessary constant is $\left(\frac{1}{2}(6)\right)^2$, or 9. For the y-terms $\left(\frac{1}{2}(-12)\right)^2$, or 36, is needed. Add the required constants to each side of the equation.

$$x^2 + 6x + 9 + y^2 - 12y + 36 = -25 + 9 + 36$$

$(x + 3)^2 + (y - 6)^2 = 20$ factor the x-terms and the y-terms

The center has coordinates $(-3, 6)$ and the radius is $\sqrt{20}$.

11 A point that lies on a curve will satisfy the equation of the curve, so the strategy is to write the equation of the circle. Then substitute the coordinates of point A for x and y in the equation and check if the equation is balanced.

$(x - h)^2 + (y - k)^2 = R^2$ equation of a circle

$(x - 4)^2 + (y + 5)^2 = \sqrt{65}^2$ substitute $h = 4, k = -5$, and $R = \sqrt{65}$

$(8 - 4)^2 + (2 + 5)^2 = \sqrt{65}^2$ substitute coordinates of A for x and y

$$4^2 + 7^2 = \sqrt{65}^2$$
$$16 + 49 = 65$$
$$65 = 65$$

The equation is balanced, so the point $A(8, 2)$ does lie on circle P.

12 Opposite sides of a parallelogram are parallel, so side $\overline{CD}$ is parallel to side $\overline{AB}$ and must have the same slope.

Start by finding the slope of $\overline{AB}$.

$$\text{slope } \overline{AB} = \frac{y_2 - y_1}{x_2 - x_1}$$
$$= \frac{1 - (-1)}{5 - 2}$$
$$= \frac{2}{3}$$

The desired line also passes through point $D(3, 4)$; $x = 3$ and $y = 4$ must satisfy the equation of $\overline{CD}$. Apply the point-slope equation of a line with slope $= \frac{2}{3}$, which represents the slope of $\overline{AB}$ using $x_1 = 3$ and $y_1 = 4$:

$$y - y_1 = m(x - x_1)$$
$$y - 4 = \frac{2}{3}(x - 3)$$

or, in slope-intercept form $y = \frac{2}{3}x + 2$.

13 We can estimate the area by filling in the ellipse with 5 rectangles. Triangles could have been used to get a more accurate estimate of the area.

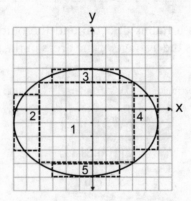

The areas of the rectangles are

    rectangle 1 area $= 7 \cdot 6 = 42$

    rectangle 2 area $= 4 \cdot 2 = 8$

    rectangle 3 area $= 5 \cdot 1 = 5$

    rectangle 4 area $= 4 \cdot 2 = 8$

    rectangle 5 area $= 5 \cdot 1 = 5$

The approximate area of the ellipse is $42 + 8 + 5 + 8 + 5 = 68$ square units.

14 The two circles are the largest that will fit inside the region, and the smallest that will fit outside the region.

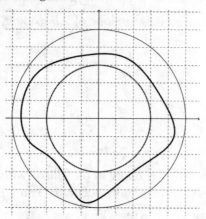

The inner circle has a radius of 3 cm and the outer circle has a radius of 5 cm. The estimate is the average of the areas of the two circles.

$$A_{estimate} = \frac{1}{2}\left(\pi R^2_{inner} + \pi R^2_{outer}\right)$$

$$= \frac{1}{2}\left(\pi \cdot 3^2 + \pi \cdot 5^2\right)$$

$$= \frac{1}{2}(106.8141)$$

$$\approx 53.4 \text{ cm}^2$$

15 a) To find the equation of a perpendicular bisector, the slope and midpoint of $\overline{SK}$ are needed.

$$\text{slope } \overline{SK} = \frac{y_2 - y_1}{x_2 - x_1}$$

$$= \frac{0 - 2}{6 - 0}$$

$$= -\frac{1}{3}$$

midpoint $\overline{SK}$ is $\dfrac{x_1 + x_2}{2}, \dfrac{y_1 + y_2}{2}$

$$\left(\frac{0 + 6}{2}, \frac{2 + 0}{2}\right)$$

$$(3, 1)$$

Perpendicular lines have negative reciprocal slopes, so the slope of the perpendicular bisector is 3, and the line must pass through the midpoint $(3, 1)$. Substitute these values in the point-slope equation of a line:

$$y - y_1 = m(x - x_1)$$
$$y - 1 = 3(x - 3)$$

or $y = 3x - 8$ in slope-intercept form.

b) An altitude will pass through the opposite vertex, so check if the coordinates of point $Y(4, 4)$ satisfy the equation of the perpendicular bisector.

$$y - 1 = 3(x - 3),$$
$$4 - 1 = 3(4 - 3)$$
$$3 = 3(1)$$
$$3 = 3$$

The equation is balanced, so the perpendicular bisector passes through the opposite vertex and is also an altitude.

3.7 SIMILAR FIGURES

PROPERTIES AND DEFINITION OF SIMILAR FIGURES

Two figures are similar if there exists a sequence of similarity transformations that maps one figure onto another. Remember, the sequence of similarity transformations often includes a dilation.

Similar figures have the following properties:

- Same shape but possibly different sizes
- All corresponding lengths proportional
- All corresponding angles congruent

The ratio of proportionality is called the *scale factor*. A similarity statement is written using the symbol $\sim$. In the accompanying figure, $\triangle ABC \sim \triangle DEF$, from the similarity statement, we have the following relationships:

$$\angle A \cong \angle D \quad \angle B \cong \angle E \quad \angle C \cong \angle F$$

$$\frac{AB}{DE} = \frac{BC}{EF} = \frac{AC}{DF} = \frac{1}{3}$$

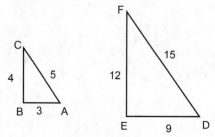

PROVING TRIANGLES SIMILAR USING THE SIMILARITY POSTULATES

- Two triangles can be proven similar using the following postulates:
  AA (angle–angle)
  SAS (side–angle–side)
  SSS (side–side–side)

Unlike congruence, pairs of corresponding sides must be in the same ratio.

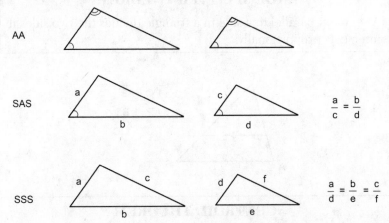

AA

SAS $\quad \dfrac{a}{c} = \dfrac{b}{d}$

SSS $\quad \dfrac{a}{d} = \dfrac{b}{e} = \dfrac{c}{f}$

PERIMETER, AREA, AND VOLUME IN SIMILAR FIGURES

Perimeter, area, and volume in similar figures and solids are related to the ratio of corresponding sides.

$\dfrac{\text{Perimeter}_A}{\text{Perimeter}_B}$	$\dfrac{\text{Area}_A}{\text{Area}_B}$	$\dfrac{\text{Volume}_A}{\text{Volume}_B}$
scale factor	(scale factor)^2	(scale factor)^3

SIMILARITY RELATIONSHIPS IN TRIANGLES

SEGMENT PARALLEL TO A SIDE THEOREM
- A segment parallel to a side of a triangle forms a triangle similar to the original triangle.
- If a segment intersects two sides of a triangle such that a triangle similar to the original is formed, the segment is parallel to the third side of the original triangle.

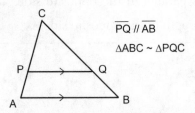

$\overline{PQ} \parallel \overline{AB}$

$\triangle ABC \sim \triangle PQC$

SIDE SPLITTER THEOREM

A segment parallel to a side in a triangle divides the two sides it intersects proportionally.

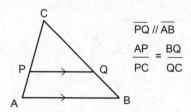

$$\overline{PQ} \,/\!/\, \overline{AB}$$

$$\frac{AP}{PC} = \frac{BQ}{QC}$$

CENTROID THEOREM

The centroid of a triangle divides each median in a 1 : 2 ratio, with the longer segment having a vertex as one of its endpoints.

In the accompanying figure, D is the midpoint of $\overline{AB}$, and E is the midpoint of $\overline{BC}$, making $\overline{AE}$ and $\overline{CD}$ medians. Their point of intersection, G, is the centroid and divides the medians in a 1 : 2 ratio.

$$\frac{EG}{AG} = \frac{1}{2} \text{ and } \frac{DG}{CG} = \frac{1}{2}$$

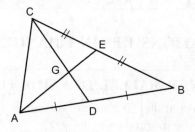

MIDSEGMENT THEOREM

A segment joining the midpoints of two sides of a triangle (a midsegment) is parallel to the opposite side, and its length is equal to $\frac{1}{2}$ the length of the opposite side.

In the accompanying figure, E is the midpoint of $\overline{AB}$, and F is the midpoint of $\overline{AC}$. $\overline{EF}$ // $\overline{BC}$ and $EF = \frac{1}{2}BC$.

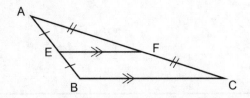

ALTITUDE TO THE HYPOTENUSE OF A RIGHT TRIANGLE THEOREM

The altitude to the hypotenuse of a right triangle forms two triangles that are similar to the original triangle.

In the accompanying figure, $\overline{CD}$ is an altitude to hypotenuse $\overline{AB}$, $\triangle BDC \sim \triangle CDA \sim \triangle BCA$.

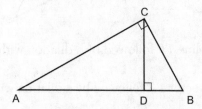

Two useful strategies for dealing with these overlapping similar triangles is to sketch the triangles separately or to make a table with sides classified as the "leg 1," "leg 2," and "hypotenuse." The altitude will be leg 1 of one of the interior triangles, and leg 2 of the other.

	Leg 1	**Leg 2**	**Hypotenuse**
Small △	$\overline{BD}$	$\overline{CD}$	$\overline{BC}$
Medium △	$\overline{CD}$	$\overline{AD}$	$\overline{AC}$
Large △	$\overline{BC}$	$\overline{AC}$	$\overline{AB}$

Practice Exercises

1 In the accompanying figure, $\triangle PQR \sim \triangle TSR$ and points P, R, and T are collinear. Which of the following transformations will map $\triangle TSR$ to $\triangle PQR$?

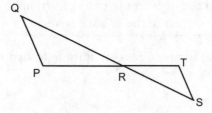

(1) reflection over line $\overline{TP}$ followed by a dilation with a scale factor of $\dfrac{QR}{TR}$ centered at Q

(2) translation by vector $\overline{TR}$ followed by a dilation with a scale factor of $\dfrac{PQ}{ST}$ centered at R

(3) rotation of $180°$ about point R followed by a dilation with a scale factor of $\dfrac{PQ}{ST}$ centered at R

(4) rotation of $180°$ about point R followed by a dilation with a scale factor of $\dfrac{QR}{TR}$ centered at R

2 $ABCDE$ and $VWXYZ$ are similar pentagons. If $BC = 4$ and $WX = 6$, what is the ratio of the area of $ABCDE$ to the area of $VWXYZ$?

(1) $\dfrac{4}{9}$ (3) $\dfrac{2}{5}$

(2) $\dfrac{2}{3}$ (4) $\sqrt{\dfrac{2}{3}}$

3 If $\triangle XYZ \sim \triangle XVW$, which of the following transformations will map $\triangle XVW$ to $\triangle XYZ$?

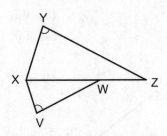

(1) reflect $\triangle XVW$ over $\overline{XW}$ and then dilate it with a center at X and scale factor of $\dfrac{XW}{WZ}$

(2) reflect $\triangle XVW$ over $\overline{XW}$ and then dilate it with a center at X and scale factor of $\dfrac{XZ}{XW}$

(3) rotate $\triangle XVW$ $180°$ about point X and then dilate it with a center at X and scale factor of $\dfrac{XW}{WZ}$

(4) rotate $\triangle XVW$ $180°$ about point X and then dilate it with a center at X and scale factor of $\dfrac{XZ}{XW}$

4 Which of the following triangles can be proven to be similar?

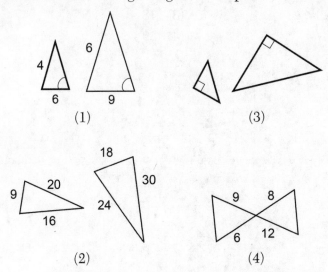

5 Given triangles FOG and BAT with $\angle G \cong \angle T$, $\angle O \cong \angle B$, $FG = 5x$, $FO = 42$, $AB = 14$, and $AT = x + 12$, find the length FG.

(1) 18   (3) 45
(2) 30   (4) 90

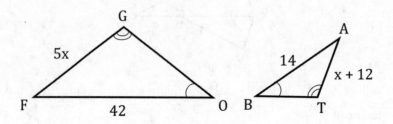

6 $\overline{UV}$ is parallel to side $\overline{ST}$ of $\triangle RST$. If $RU = x - 6$, $US = x$, $RV = x$, and $VT = 2x - 9$, what is the value of x?

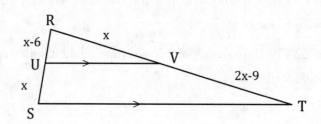

(1) 3   (3) 18
(2) 11   (4) 39

7 In the figure below, D lies on $\overline{AB}$ and E lies on $\overline{AC}$. If $AD = 3$, $BD = x + 3$, $DE = 5$, and $BC = 2x + 2$, find the value of x that would let you conclude that $\overline{DE} \mathbin{/\mkern-5mu/} \overline{BC}$.

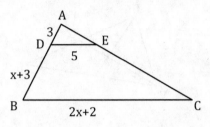

(1) 2 (3) 18

(2) 9 (4) 24

8 Mollie wants to rearrange the furniture in her living room. She measures the rectangular room to be 16 ft by 12 ft and makes a scaled drawing of the room that measures 10 in by 7.5 in. She wants to make scaled paper cutouts to represent her furniture so she can determine her favorite arrangement without having to move the furniture. If her sofa is rectangular and measures 6 ft by 2 ft, what should be the dimensions of the cutout?

(1) $1\frac{1}{3}$ in $\times$ 1 in (3) $3\frac{3}{4}$ in $\times 1\frac{1}{4}$ in

(2) $2\frac{3}{4}$ in $\times$ 1 in (4) 9.6 in $\times$ 3.2 in

9 Given: $\overline{LSI}$ and $\overline{FSP}$ intersecting at S, $\overline{FL} \mathbin{/\mkern-5mu/} \overline{IP}$, $IP = 4$ and $FL = 6$

(1) Explain why the two triangles formed are similar.

(2) Find the ratio $\dfrac{PS}{FS}$.

(3) State a specific similarity transformation that will map the smaller triangle onto the larger.

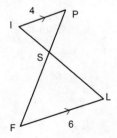

10 In $\triangle VTP$, F is the midpoint of $\overline{VP}$, G is the midpoint of $\overline{PT}$, and $\overline{VG}$ and $\overline{FT}$ intersect at K. If $VK = 5x + 5$ and $KG = 2x + 8$, find the length VG.

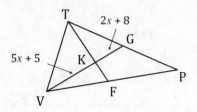

11 The midpoints of $\overline{PG}$, $\overline{PI}$, and $\overline{GI}$ are C, O, and W, respectively. If $OW = 7$, $CW = 5$, and $OC = 6$, find the perimeter of $\triangle PIG$.

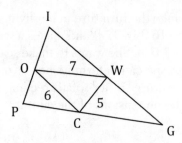

12 In the figure below, $\angle ABC$ and $\angle BDC$ are right angles, $CB = 6$, and $AB = 12$. Find the length of $\overline{BD}$. Express your answer in simplest radical form.

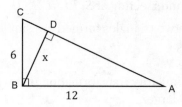

13 Rectangle *AEFG* is the image of rectangle *ABCD* after the transformation clockwise $R_{90°,A} \circ D_{\frac{2}{3},A}$. If $GA = 8$ and $DE = 12$, what is the perimeter of *ABCD*?

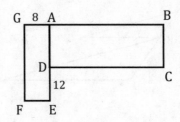

14 *ABCDEF* and *ARSTUV* are both regular hexagons and *R* is the midpoint of $\overline{AB}$. Name a sequence of transformations that will map *ARSTUV* to *ABCDEF*.

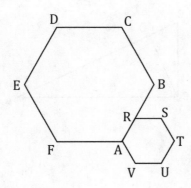

15 Given: $\overline{RUT}$, $\overline{SVT}$, and $\overline{UV} \parallel \overline{RS}$

   Prove: $\dfrac{UR}{TU} = \dfrac{VS}{TV}$

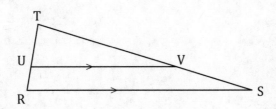

16 Given: $\overline{ABD}$, $\angle C \cong \angle E$, $\angle DEB \cong \angle EBC$

Prove: $\triangle ABC \sim \triangle BDE$

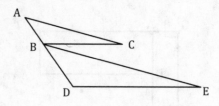

17 Prove the Pythagorean theorem.

Given: right triangle $\triangle BCA$ with a right angle at C, altitude $\overline{CD}$ drawn to hypotenuse $\overline{AB}$

Prove: $(AC)^2 + (BC)^2 = (AB)^2$

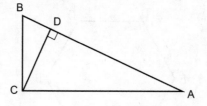

Solutions

1 We want to map the smaller triangle to the larger, so the scale factor will be greater than 1. The correct scale factor among the choices is $\dfrac{PQ}{ST}$. A rotation of $180°$ will rotate $\triangle TSR$ to $\triangle PQR$. The correct sequence is: a rotation of $180°$ centered at R, followed by a dilation by a scale factor of $\dfrac{PQ}{ST}$ centered at R. Note that the order of the sequence does not matter in this situation.

The correct choice is (**3**).

2 The ratio of the areas of similar figures is equal to the ratio of lengths of corresponding sides squared. $\overline{BC}$ and $\overline{WX}$ are corresponding sides, and are in a ratio of $\dfrac{4}{6}$, or $\dfrac{2}{3}$.

$$\text{ratio of areas} = \text{ratio of sides}^2$$

$$= \left(\frac{2}{3}\right)^2$$

$$= \frac{4}{9}$$

The correct choice is (**1**).

3 A reflection over $\overline{XW}$ will map $\overline{XV}$ onto $\overline{XZ}$. Then a dilation centered about X with a scale factor of $\dfrac{XZ}{XW}$ will match the sizes.

The correct choice is (**2**).

4 The triangle similarity postulates are **AA**, **SAS**, and **SSS**, where each S represents the *ratio* of a pair of corresponding sides. Examine each of the choices.

Choice 1: The two pairs of corresponding sides are in a 2:3 ratio, but the congruent angle is not the included angle.

Choice 2: Corresponding sides are not in the same ratio, $\dfrac{9}{18} \neq \dfrac{16}{24}$.

Choice 3: Lengths of the sides are not given.

Choice 4: 2 pairs of corresponding sides are in the same ratio, $\frac{6}{8} = \frac{9}{12}$, and the congruent vertical angles are the included angles. The triangles are similar by SAS.

The correct choice is (**4**).

5 The two triangles are similar by AA, and corresponding sides of similar triangles are in the same ratio. The similarity statement can be written as $\triangle FOG \sim \triangle ABT$, so the following proportion can be written:

$$\frac{FG}{AT} = \frac{FO}{AB}$$

$$\frac{5x}{x+12} = \frac{42}{14}$$

$$14(5x) = 42(x+12) \quad \text{cross products are equal}$$

$$70x = 42x + 504$$

$$28x = 504$$

$$x = 18 \qquad \text{substitute } x = 18$$

$$FG = 5(18)$$

$$= 90$$

The correct choice is (**4**).

6 When a segment is drawn parallel to a side of a triangle, the two triangles are similar, and the side splitter theorem states that the sides are divided proportionally.

$$\frac{RU}{US} = \frac{RV}{VT} \qquad \text{side splitter theorem}$$

$$\frac{x-6}{x} = \frac{x}{2x-9}$$

$$(2x - 9)(x - 6) = x^2 \quad \text{cross products are equal}$$

$$2x^2 - 21x + 54 = x^2$$

$$x^2 - 21x + 54 = 0 \quad \text{solve by moving all terms to one side and factoring}$$

$$(x - 18)(x - 3) = 0 \quad \text{factor the trinomial}$$

$$x - 18 = 0 \quad x - 3 = 0 \quad \text{zero product theorem}$$

$$x = 18 \qquad x = 3$$

The two solutions of the quadratic are *possible* solutions to the problem. Eliminate any results that lead to negative lengths or any other impossible situations. The solution $x = 3$ would result in a negative length for RU and VT, so we eliminate this result, and the value of x is 18.

The correct choice is **(3)**.

7 From the original figure, $\overline{DE} \mathbin{/\mkern-5mu/} \overline{BC}$ when $\angle CBA \cong \angle EDA$ (corresponding angles are congruent). The triangles $\triangle ADE$ and $\triangle ABC$ are similar by AA, and corresponding sides will be proportional. Sketch the two triangles separately to help identify the corresponding sides. Note that $AB = AD + DB$.

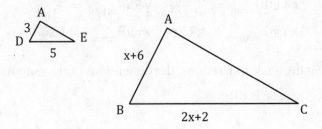

$$\frac{AD}{AB} = \frac{DE}{BC} \qquad \text{corresponding sides are proportional}$$

$$\frac{3}{x + 6} = \frac{5}{2x + 2}$$

$$3(2x + 2) = 5(x + 6) \quad \text{cross products are equal}$$

$$6x + 6 = 5x + 30$$

$$x = 24$$

The correct choice is **(4)**.

8 First convert feet to inches so all units are consistent.

$$\text{Length of room} \quad 16 \text{ ft} \times \frac{12 \text{ in}}{\text{ft}} = 192 \text{ in}$$

$$\text{Length of sofa} \quad 6 \text{ ft} \times \frac{12 \text{ in}}{\text{ft}} = 72 \text{ in}$$

$$\text{Width of sofa} \quad 2 \text{ ft} \times \frac{12 \text{ in}}{\text{ft}} = 24 \text{ in}$$

The scaled drawing is similar to the actual room, so the ratio of corresponding sides is

$$\frac{\text{length of drawing}}{\text{length of room}} = \frac{10}{192}$$

Apply this scale factor to the length and width of the sofa to determine the dimensions of the cutout of the sofa.

$$\frac{\text{length}_{\text{cutout}}}{\text{length}_{\text{actual}}} = \frac{10}{192} \qquad \frac{\text{width}_{\text{cutout}}}{\text{width}_{\text{actual}}} = \frac{10}{192}$$

$$\frac{\text{length}_{\text{cutout}}}{72 \text{ in}} = \frac{10}{192} \qquad \frac{\text{width}_{\text{cutout}}}{24 \text{ in}} = \frac{10}{192}$$

$$\text{length}_{\text{cutout}} = \frac{72 \cdot 10}{192} \qquad \text{width}_{\text{cutout}} = \frac{24 \cdot 10}{192}$$

$$\text{length}_{\text{cutout}} = 3.75 \text{ in} \qquad \text{width}_{\text{cutout}} = 1.25 \text{ in}$$

Converting decimals to fractions, the dimensions of the cutout are $3\frac{3}{4} \times 1\frac{1}{4}$. The correct choice is **(3)**.

9 a) Given $\overline{FL} \parallel \overline{IP}$, we know $\angle F \cong \angle P$ and $\angle L \cong \angle I$ because they are the alternate interior angles formed by parallel lines and a transversal. Therefore, $\triangle SIP \sim \triangle SLF$ by AA. (Note the correct order of vertices in the similarity statement.)

b) Corresponding parts are in the same ratio, so

$$\frac{PS}{FS} = \frac{IP}{LF}$$
$$= \frac{4}{6}$$
$$= \frac{2}{3}$$

c) We have to use the reciprocal of the ratio in part (b) because we want to enlarge the smaller triangle and not reduce it. A rotation of $180°$ about point S will align $\overline{SP}$ with $\overline{FS}$ and $\overline{IS}$ with $\overline{SL}$. A dilation with center at S and a scale factor of $\frac{3}{2}$ will then map $\triangle SIP$ to $\triangle SLF$.

10 $\overline{VG}$ and $\overline{TF}$ are medians that intersect at centroid K. The centroid divides $\overline{VG}$ in a $1:2$ ratio, so

$$VK = 2KG$$
$$5x + 5 = 2(2x + 8)$$
$$5x + 5 = 4x + 16$$
$$5x = 4x + 11$$
$$x = 11$$

VG is equal to the sum of VK and KG, which can be evaluated at $x = 11$.

$$VG = VK + KG$$
$$= 5x + 5 + 2x + 8$$
$$= 7x + 13$$
$$= 7(11) + 13$$
$$= 90$$

11 $\overline{CO}$, $\overline{OW}$, and $\overline{WC}$ are all midsegments, and their lengths are $\frac{1}{2}$ the length of the opposite side of $\triangle PIG$. Therefore,

$$PI = 2 \cdot WC$$
$$= 2 \cdot 5$$
$$= 10$$

$$IG = 2 \cdot OC$$
$$= 2 \cdot 6$$
$$= 12$$

$$GP = 2 \cdot WO$$
$$= 2 \cdot 7$$
$$= 14$$

The perimeter of $\triangle PIG = PI + IG + GP$
$$= 10 + 12 + 14$$
$$= 36$$

12 An altitude to the hypotenuse of a right triangle forms 3 similar triangles. The three similar triangles are $\triangle CDB$, $\triangle BDA$, and $\triangle CBA$. Fill in the table of side lengths, and look for two pairs of corresponding sides that can be used to write a proportion involving the unknown length x.

	Leg 1	**Leg 2**	**Hypotenuse**
Small $\triangle CDB$		x	6
Medium $\triangle BDA$	x		12
Large $\triangle CBA$	6	12	

We don't have a full set of four corresponding sides, so we need to use the Pythagorean theorem on the large triangle to find the hypotenuse AC.

$$BC^2 + AB^2 = AC^2$$
$$6^2 + 12^2 = AC^2$$
$$AC^2 = 180$$
$$AC = \sqrt{180}$$
$$= 6\sqrt{5}$$

Now the table indicates two pairs of corresponding parts can be used to write a proportion.

	Leg 1	Leg 2	Hypotenuse
Small $\triangle CDB$		x	6
Medium $\triangle BDA$	x		12
Large $\triangle CBA$	6	12	$6\sqrt{5}$

$$\frac{x}{12} = \frac{6}{6\sqrt{5}}$$

$6x\sqrt{5} = 72 \qquad$ cross products are equal

$$x = \frac{72}{6\sqrt{5}}$$

$$= \frac{12}{\sqrt{5}}$$

$$= \frac{12}{\sqrt{5}} \cdot \frac{\sqrt{5}}{\sqrt{5}} \qquad \text{rationalize the denominator}$$

$$= \frac{12\sqrt{5}}{5}$$

13 The dilation tells us that the ratio of corresponding parts between rectangles $AEFG$ and $ABCD$ is 2:3. GA and AD are corresponding parts, so the following proportion can be written:

$$\frac{GA}{AD} = \frac{2}{3}$$

$$\frac{8}{AD} = \frac{2}{3}$$

$2AD = 24 \qquad$ cross products are equal

$AD = 12$

$AE = AD + DE \qquad$ segment addition

$AE = 12 + 12$

$ = 24$

Now we can find AB by writing a proportion involving corresponding sides AE and AB.

$$\frac{AE}{AB} = \frac{2}{3}$$
$$3AE = 2AB \quad \text{cross products are equal}$$
$$3(24) = 2AB$$
$$72 = 2AB$$
$$AB = 36$$

Since $ABCD$ is a rectangle, opposite sides are congruent, and the perimeter is equal to

$$\text{perimeter}_{ABCD} = 2AB + 2BC$$
$$= 2(36) + 2(12)$$
$$= 72 + 24$$
$$= 96$$

14 A dilation and a reflection are required to map $ARSTUV$ to $ABCDEF$. Since R is a midpoint, we know $AR = \frac{1}{2}AB$, and a dilation with a scale factor of 2 centered at A is needed to map $\overline{AR}$ to $\overline{AB}$. The second transformation needed is a reflection over line $\overleftrightarrow{AB}$.

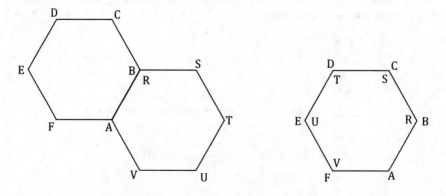

The transformation is a dilation centered at A with a scale factor of 2, followed by a reflection over $\overline{AB}$. Alternatively, the order could be reversed.

15   The strategy is to prove $\triangle TUV \sim \triangle TRS$, which leads to the proportion of corresponding sides $\dfrac{TR}{TU} = \dfrac{TS}{TV}$. Algebraic rearranging gives the final result.

Statement	Reason
1. $\overline{RUT}$, $\overline{SVT}$, and $\overline{UV}$ // $\overline{RS}$	1. Given
2. $\angle TUV \cong \angle R$ and $\quad\;\; \angle TVU \cong \angle S$	2. Corresponding angles formed by parallel lines are congruent
3. $\triangle TUV \sim \triangle TRS$	3. AA
4. $\dfrac{TR}{TU} = \dfrac{TS}{TV}$	4. Corresponding sides in similar triangles are proportional
5. $TR = TU + UR$ $\quad\;\; TS = TV + VS$	5. Partition
6. $\dfrac{TU + UR}{TU} = \dfrac{TV + VS}{TV}$	6. Substitution
7. $1 + \dfrac{UR}{TV} = 1 + \dfrac{VS}{TV}$	7. Division
8. $\dfrac{UR}{TU} = \dfrac{VS}{TV}$	8. Subtraction

16   The strategy is to show that $\overline{BC}$ // $\overline{DE}$ and then to use the corresponding angle relationship to show $\angle ABC \cong \angle BDE$.

Statement	Reason
1. $\overline{ABD}$, $\angle C \cong \angle E$, $\quad\;\; \angle DEB \cong \angle EBC$	1. Given
2. $\overline{BC}$ // $\overline{DE}$	2. Two lines are parallel if the alternate interior angles formed by a transversal are congruent
3. $\angle ABC \cong \angle BDE$	3. The corresponding angles formed by two parallel lines are congruent
4. $\triangle ABC \sim \triangle BDE$	4. AA

17 The strategy is to use the theorem that an altitude to the hypotenuse of a right triangle forms three similar triangles. Two proportions from these triangles can be written and combine to yield the Pythagorean theorem.

Statement	Reason
1. Right triangle $\triangle ABC$ with a right angle at C, altitude $\overline{CD}$ drawn to hypotenuse $\overline{AC}$	1. Given
2. $\triangle BDC \sim \triangle CDA \sim \triangle BCA$	2. The altitude to the hypotenuse of a right triangle forms 3 similar right triangles
3. $\dfrac{AB}{AC} = \dfrac{AC}{AD}, \dfrac{AB}{BC} = \dfrac{BC}{BD}$	3. Corresponding sides in similar triangles are proportional
4. $(AC)^2 = AB \cdot AD,$ $(BC)^2 = AB \cdot BD$	4. The cross products of a proportion are equal
5. $(AC)^2 + (BC)^2 = AB \cdot AD + AB \cdot BD$	5. Addition property
6. $(AC)^2 + (BC)^2 = AB(AD + BD)$	6. Factor
7. $(AC)^2 + (BC)^2 = AB(AB)$	7. Partition property
8. $(AC)^2 + (BC)^2 = (AB)^2$	8. Simplify

3.8 TRIGONOMETRY

DEFINITION OF SIDES IN A RIGHT TRIANGLE RELATIVE TO AN ANGLE

- Hypotenuse—the side across from the right angle
- Opposite—the side across from the specified acute angle
- Adjacent—the side included between the right angle and the specified angle

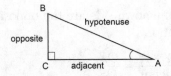   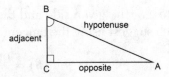

a) Opposite, adjacent, and   b) Opposite, adjacent, and
hypotenuse relative to $\angle A$ hypotenuse relative to $\angle B$

TRIGONOMETRIC RATIOS

Ratio	Abbreviation	Definition
Sine	Sin	$\dfrac{\text{opposite}}{\text{hypotenuse}}$
Cosine	Cos	$\dfrac{\text{adjacent}}{\text{hypotenuse}}$
Tangent	Tan	$\dfrac{\text{opposite}}{\text{adjacent}}$

The ratios sine, cosine, and tangent will have fixed values for a given acute angle in a right triangle. This is because any two right triangles with a pair of congruent acute angles are similar, and corresponding parts of similar triangles are in the same ratio. These three trigonometric ratios can be used to find a missing side of a triangle given one angle and side by setting up the correct proportion. Be sure your calculator is in degree mode if the angle is given in degrees.

Example

$\triangle SOX$ has a right angle at O. The measure of angle $S = 72°$ and $XO = 12$. Find the length OS to the nearest tenth.

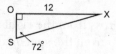

Solution:

Relative to $\angle S$, the given side and the unknown side are the opposite and adjacent, which suggest using the tangent ratio.

$$\tan(S) = \frac{\text{opposite}}{\text{adjacent}} = \frac{OX}{OS}$$

$$\tan(72°) = \frac{12}{OS}$$

$$OS = \frac{12}{\tan(72°)}$$

$$OS = \frac{12}{3.07768} = 3.89903$$

$$= 3.9$$

INVERSE TRIGONOMETRIC RATIOS

The inverse of a function "undoes" the original function. For example, $f(x) = \sqrt{x}$ and $g(x) = x^2$ are inverse functions. If $x = 5$, then $f(g(5)) = \sqrt{5^2} = 5$. The trigonometric functions have the following inverse functions:

$\sin(x)$	$\arcsin(x)$
$\cos(x)$	$\arccos(x)$
$\tan(x)$	$\arctan(x)$

- These three inverse functions are often abbreviated $\sin^{-1}(x)$, $\cos^{-1}(x)$, and $\tan^{-1}(x)$. They can be used to find the measure of an angle given a ratio of any two sides in the triangle.

Example

Using the accompanying figure, find the measure of angles M and N.

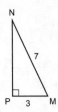

Solution:

$$\cos(M) = \frac{3}{7}$$
$$m\angle M = \cos^{-1}\!\left(\frac{3}{7}\right)$$
$$= 64.6°$$
$$\sin(N) = \frac{3}{7}$$
$$m\angle N = \sin^{-1}\!\left(\frac{3}{7}\right)$$
$$= 25.4°$$

COFUNCTIONS

Cofunction Relationship

The sine of an angle is equal to the cosine of its complement: $\sin(A) = \cos(90 - A)$.

The cosine of an angle is equal to the sine of its complement: $\cos(A) = \sin(90 - A)$.

$$\sin(A) = \cos(B) = \cos(90 - A) = \frac{BC}{AB}$$
$$\sin(B) = \cos(A) = \sin(90 - A) = \frac{AC}{AB}$$

In other words, the cofunction relationship states that if a sine is equal to a cosine, then the two angles must sum to 90°.

Example

Given $\sin(2x + 4) = \cos(3x + 21)$. Solve for x.

Solution:

The cofunctions are equal, so the angles must be complementary.

$$
\begin{aligned}
2x + 4 + 3x + 21 &= 90 \\
5x + 25 &= 90 \\
5x &= 65 \\
x &= 13
\end{aligned}
$$

MODELING WITH TRIGONOMETRY

- *Angle of elevation*—The angle formed by the horizontal and the line directed upward to an object
- *Angle of depression*—The angle formed by the horizontal and the line directed downward to an object

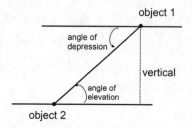

Strategy for trigonometry modeling problems:

- Start with a detailed sketch.
- Identify the triangles formed and the relevant trigonometric ratios.
- Consider parallel lines and other relationships if you are still missing parts. Notice that the angle of elevation and angle of depression are congruent alternate interior angles.
- Consider working with more than one triangle, especially if the problem involves an object that moves from one position to another.

Practice Exercises

1 In $\triangle ABC$, $m\angle B = 90°$, $m\angle C = 51°$, and $BC = 13$. What is the length AB?

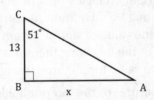

(1) 16.1

(3) 18.2

(2) 16.7

(4) 20.7

2 $\triangle JPL$ is a right triangle with a right angle at L. If $m\angle J = 19°$ and $\overline{LJ} = 21$, what is the length of $\overline{PJ}$?

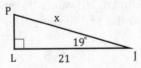

(1) 21.5

(3) 61.0

(2) 22.2

(4) 64.5

3 A 24 ft ladder is leaning against a wall. The base of the ladder is 5 ft from the house. What angle does the ladder make with the ground?

(1) 12°   (3) 67°

(2) 65°   (4) 78°

4 In right triangle PQR, m$\angle Q = 90°$. If $\sin(P) = 4x^2 + 5x + 2$ and $\cos(R) = 5x + 3$, what is the measure of $\angle P$?

(1) 5°   (3) 25°

(2) 8°   (4) 30°

5 John constructs three right triangles of different sizes, each having angles that measure 30°, 60°, and 90°. He then measures the side lengths of each triangle and calculates the value of $\sin(60°)$ from the side lengths using each of the triangles. Which of the following theorems would best explain why the value he calculates is the same for all three triangles?

(1) The sum of the measures of the interior angles of a triangle equals 180°.

(2) Corresponding parts of congruent triangles are congruent.

(3) Corresponding sides of similar triangles are proportional.

(4) The longest side of a triangle is always opposite the largest angle.

6 The side of Rayquan's house is 30 ft long. Rayquan is planning to install a gutter along the roof as shown. He wants the angle of depression to be 3° so the water will drain out at the end of the gutter. How long should the gutter be?

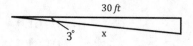

(1) 31 ft 7 in   (3) 30 ft 4 in

(2) 31 ft 2 in   (4) 30 ft 0.5 in

7 In $\triangle ABC$, $m\angle B = 90°$, $\tan(A) = m$, and $\sin(C) = \dfrac{1}{p}$. What is the value of $\cos(C)$?

(1) $\dfrac{m}{p}$

(3) $\dfrac{p}{m}$

(2) $m \cdot p$

(4) $m + p$

8 Jack is watching the launch of a rocket from a viewing area 4,500 feet from the launch pad. At 15 seconds after launch, he measures a 67° angle of elevation from the ground to the rocket. What is the average speed of the rocket during the first 15 seconds of its flight? Assume the rocket travels upward, perpendicular to the ground, and give your answer to the nearest foot per second.

9 Rick is flying a kite in the park. He holds the spool of string 3 ft above the ground, and lets out 200 ft of string. The kite is initially flying with a 72° angle of elevation.

(1) What is the altitude of the kite?

(2) The wind shifts and the angle of elevation altitude decreases to 52°. How many feet does the kite drop?

10 Frank is standing at the top of a water tower and looks down at the town below. Frank measures an angle of depression of 52° to his house, and he knows the water tower is 120 ft tall. How far along the ground is the tower from his house? Round to the nearest foot.

11 A lighthouse sits on a 100 ft cliff. An observer in a boat in the harbor notes an angle of elevation to the base of the cliff of 28°, and an angle of elevation to the top of the lighthouse of 44°. What is the height of the lighthouse, rounded to the nearest foot?

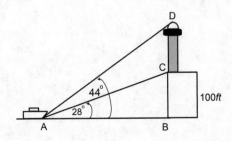

Solutions

1 Relative to $\angle C$, $\overline{AB}$ is the opposite and $\overline{BC}$ is the adjacent, so the tangent ratio applies.

$$\text{tangent} = \frac{\text{opposite}}{\text{adjacent}}$$

$$\tan(C) = \frac{AB}{BC}$$

$$\tan(51°) = \frac{x}{13}$$

$$x = 13\tan(51°)$$

$$= 16.0536$$

$$= 16.1$$

The correct choice is (**1**).

2 Relative to $\angle J$, $\overline{LJ}$ is the adjacent and $\overline{JP}$ is the hypotenuse. The cosine ratio applies here.

$$\text{cosine} = \frac{\text{adjacent}}{\text{hypotenuse}}$$

$$\cos(J) = \frac{LJ}{JP}$$

$$\cos(19°) = \frac{21}{x}$$

$$x = \frac{21}{\cos(19°)}$$

$$= 22.2100$$

$$= 22.2$$

The correct choice is (**2**).

3 The ladder, wall, and ground form a right triangle as shown.

Relative to the desired angle, the 5 ft distance is the adjacent and the 24 ft length is the hypotenuse. Adjacent and hypotenuse suggest using the cosine. Since we are looking for the angle, apply the inverse cosine function.

$$\cos(x) = \frac{\text{adjacent}}{\text{hypotenuse}}$$

$$\cos(x) = \frac{5}{24}$$

$$x = \cos^{-1}\left(\frac{5}{24}\right)$$

$$= 77.9753°$$

$$= 78°$$

The correct choice is **(4)**.

4 The sine of an angle and the cosine of its complement are always equal. Since $\triangle PQR$ is a right triangle with the right angle at Q, angles P and R must be complementary. Setting $\sin(P)$ and $\cos(R)$ equal, we get:

$$\sin(P) = \cos(R)$$

$$4x^2 + 5x + 2 = 5x + 3$$

$$4x^2 = 1$$

$$x^2 = \frac{1}{4}$$

$$x = \pm\frac{1}{2}$$

Substituting $x = -\frac{1}{2}$ into the expression for $\sin(P)$, we get

$$\sin(P) = 4x^2 + 5x + 2$$

$$= 4\left(-\frac{1}{2}\right)^2 + 5\left(-\frac{1}{2}\right) + 2$$

$$= 1 - 2\frac{1}{2} + 2$$

$$= \frac{1}{2}$$

Now substitute $x = \dfrac{1}{2}$.

$$\sin(P) = 4\left(\frac{1}{2}\right)^2 + 5\left(\frac{1}{2}\right) + 2$$
$$= 1 + 2\frac{1}{2} + 2$$
$$= 5\frac{1}{2}$$

The sine function cannot be greater than 1, so $x = \dfrac{1}{2}$ is not a solution. Using the solution $x = -\dfrac{1}{2}$ and $\sin(P) = \dfrac{1}{2}$, find $m\angle P$ using the inverse sine function.

$$\sin(P) = \frac{1}{2}$$
$$m\angle P = \sin^{-1}\left(\frac{1}{2}\right)$$
$$m\angle P = 30°$$

The correct choice is **(4)**.

5 The three triangles are all similar by the angle–angle theorem. The sine ratio is equal to the $\dfrac{\text{opposite}}{\text{hypotenuse}}$. Since all pairs of corresponding sides in similar triangles are proportional, the ratio $\dfrac{\text{opposite}}{\text{hypotenuse}}$ will be equal to the same value in all three triangles.

The correct choice is **(3)**.

6 Relative to the 3° angle, the horizontal distance is the adjacent and the gutter is the hypotenuse, so we can apply the cosine ratio.

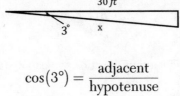

$$\cos(3°) = \frac{\text{adjacent}}{\text{hypotenuse}}$$

$$\cos(3°) = \frac{30\,\text{ft}}{x}$$

$$x = \frac{30\,\text{ft}}{\cos(3°)}$$

$$= 30.04117\,\text{ft}$$

To convert to feet and inches, multiply the decimal part of the number by 12.

$0.4117\,(12) = 0.49405\,\text{in}$

The gutter length should be 30 ft 0.49405 in. Rounded to the nearest 0.1 inch, the length is 30 ft 0.5 in.

The correct choice is **(4)**.

7 Making a sketch of $\triangle ABC$, we can fill in the lengths of AB, BC, and AC as shown in the figure.

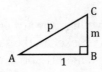

The tangent is equal to $\dfrac{\text{opposite}}{\text{adjacent}}$, so

$$\tan(A) = m$$
$$\frac{BC}{AB} = \frac{m}{1}$$

Since we are only concerned with ratios, we can assume $BC = m$ and $AB = 1$.

The second ratio states

$$\sin(C) = \frac{1}{p}$$
$$\frac{AB}{AC} = \frac{1}{p}$$

Since we assumed $AB = 1$, we can assume $AC = p$. $\text{Cos}(C)$ can now be calculated.

$$\cos(C) = \frac{\text{adjacent}}{\text{hypotenuse}}$$
$$= \frac{BC}{AC}$$
$$= \frac{m}{p}$$

The correct choice is (**1**).

8 The path of the rocket traveling upward forms a right triangle with the ground. The distance to the rocket is the adjacent, and the height of the rocket is the opposite, so the tangent ratio can be applied.

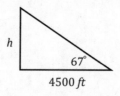

$$\tan(67°) = \frac{h}{4,500}$$

$$h = 4,500\tan(67°)$$

$$= 4,500(2.3558)$$

$$= 10,601.33 \text{ ft}$$

To find the speed, apply the relationship $\text{rate} = \frac{\text{distance}}{\text{time}}$.

$$\text{speed} = \frac{10,601.33\,\text{ft}}{15\,\text{sec}}$$

$$= 706.73\,\text{ft/sec}$$

$$= 707\,\text{ft/sec}$$

9 First make a sketch. A right triangle is formed by the kite string, the altitude measured from the spool (x), and the horizontal distance to the kite. The total altitude will be $(x + 3)$ ft. To find how many feet the kite drops, calculate the altitude at the two angles of elevation, and then find the difference.

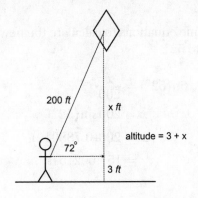

a) Relative to the angle of elevation, the 200 ft kite string is the hypotenuse, and the altitude measured from the spool is opposite. The opposite and hypotenuse suggest the sine ratio. Find x using the following:

$$\sin(72°) = \frac{\text{opposite}}{\text{hypotenuse}}$$

$$\sin(72°) = \frac{x}{200}$$

$$x = 200\sin(72°)$$

$$x = 200(0.951056)$$

$$= 190.2112\,\text{ft}$$

Next, calculate the altitude from x.

$$\text{Altitude} = 3\,\text{ft} + x\,\text{ft}$$

$$= 3\,\text{ft} + 190.21123\,\text{ft}$$

$$= 193.21123\,\text{ft}$$

$$= 193\,\text{ft}$$

b) We can use the same equations to calculate the new altitude if the angle of elevation decreases to 52°.

$$\sin(52°) = \frac{x}{200}$$

$$x = 200\sin(52°)$$

$$x = 200(0.788010)$$

$$= 157.60215\,\text{ft}$$

$$\text{Altitude} = 3\,\text{ft} + x\,\text{ft}$$

$$= 3 + 157.60215\,\text{ft}$$

$$= 160.60215\,\text{ft}$$

The change in altitude is

$$193.21123 \text{ ft} - 160.60215 \text{ ft}$$
$$= 32.6090 \text{ ft}$$
$$= 33 \text{ ft}$$

The kite drops 33 ft.

10 Start with a sketch, and fill in the appropriate dimensions, horizontals, and verticals.

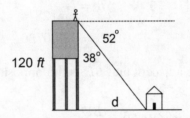

The angle of depression is 52°. It is easier in the problem to find the angle complementary to the angle of depression and work with that triangle. The complementary angle is 38°. The height of the tower is adjacent to the 38° angle, and the distance along the ground is opposite from the 38° angle, so we use the tangent ratio.

$$\tan(38°) = \frac{d}{120 \, \text{ft}}$$
$$d = 120 \, \text{ft} \cdot \tan(38°)$$
$$d = 93.75 \, \text{ft}$$

The house is 94 ft from the water tower.

11 First find length AB using $\triangle ABC$; then find length BD using $\triangle ABD$.

In $\triangle ABC$, BC is opposite $\angle A$, and AB is the adjacent, so we can use the tangent.

$$\tan = \frac{\text{opposite}}{\text{adjacent}}$$

$$\tan(\angle BAC) = \frac{BC}{AB}$$

$$\tan(28°) = \frac{100}{AB}$$

$$AB = \frac{100}{\tan(28°)}$$

$$AB = 188.0726$$

In $\triangle ABD$, AB is the adjacent and BD is the opposite, so we use the tangent again.

$$\tan(\angle BAD) = \frac{BD}{AD}$$

$$\tan(44°) = \frac{BD}{188.0726}$$

$$BD = 188.0726\tan(44°)$$

$$BD = 181.6196$$

The height of the lighthouse, DC, is found by subtracting the height of the cliff from BD.

$$DC = BD - BC$$

$$DC = 181.6196 - 100$$

$$DC = 81.6196$$

The height of the lighthouse is 82 ft.

3.9 PARALLELOGRAMS

PROPERTIES OF PARALLELOGRAMS

A **parallelogram** is a quadrilateral whose opposite sides are parallel. All parallelograms have the following properties:

1. Opposite sides are congruent.

2. Opposite angles are congruent.

3. Adjacent angles are supplementary.

4. The diagonals bisect each other.

5. The two diagonals each divide the parallelogram into two congruent triangles.

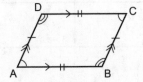 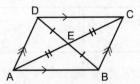

Congruent Sides:
$AB = CD, BC = AD$

Congruent angles:
$\angle A \cong \angle C, \angle B \cong \angle D$

Supplementary angles:
$\angle A$ and $\angle B$, $\angle B$ and $\angle C$
$\angle C$ and $\angle D$, $\angle D$ and $\angle A$

Diagonals Bisect Each Other:
$AE = EC, BE = ED$

Diagonals Form Congruent Triangles:
$\triangle BAD \cong \triangle DCB, \triangle ADC \cong \triangle CBA$

SPECIAL PARALLELOGRAMS

Rectangle—a parallelogram with right angles
Rhombus—a parallelogram with consecutive congruent sides
Square—a parallelogram with right angles and consecutive congruent sides

Rectangles, rhombuses, and squares have all the properties of parallelograms plus the following properties:

	Rectangle	**Rhombus**	**Square**
4 right angles	✓		✓
Congruent diagonals	✓		✓
4 congruent sides		✓	✓
Perpendicular diagonals		✓	✓
Diagonals bisect the angles		✓	✓

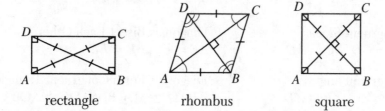

rectangle   rhombus   square

TRAPEZOIDS

A *trapezoid* is a quadrilateral with exactly one pair of parallel sides. A trapezoid with one pair of parallel sides is shown in the accompanying figure. The parallel sides are called the bases and the nonparallel sides are called the legs. The same side interior angles formed with the parallel bases are supplementary.

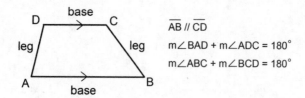

$\overline{AB} \parallel \overline{CD}$
$m\angle BAD + m\angle ADC = 180°$
$m\angle ABC + m\angle BCD = 180°$

CLASSIFYING PARALLELOGRAMS AND TRAPEZOIDS

The following table lists the properties sufficient to prove a quadrilateral is a parallelogram, rectangle, rhombus, square, or trapezoid.

Classification	What's Needed to Classify the Figure
Trapezoid	One pair of parallel sides
Parallelogram	Any *one* of the following: • Two pairs of opposite sides parallel • Two pairs of opposite sides congruent • Two pairs of opposite angles congruent • Consecutive angles supplementary • Diagonals that bisect each other • One pair of sides congruent *and* parallel
Rectangle	Any one property from the parallelogram list, plus any one of the following: • One right angle • Diagonals that are congruent
Rhombus	Any one property from the parallelogram list, plus any one of the following: • Diagonals are perpendicular • One pair of consecutive sides is congruent • A diagonal bisects one of the angles
Square	Any one property from the parallelogram list *plus* Any one property from the rectangle list *plus* Any one property from the rhombus list

The parallelogram properties needed for a proof are illustrated in the following sketches.

Property of Parallelogram	What It Looks Like
Two pairs of opposite sides are parallel.	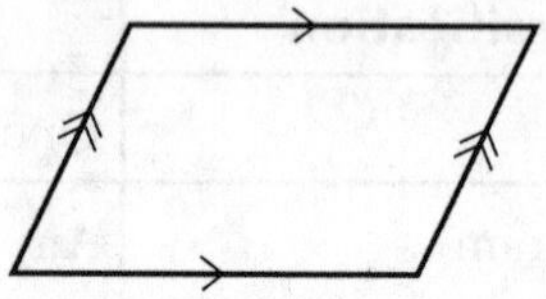
Two pairs of opposite sides are congruent.	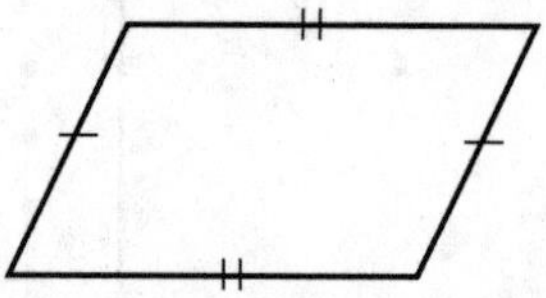
One pair of parallel sides is congruent and parallel.	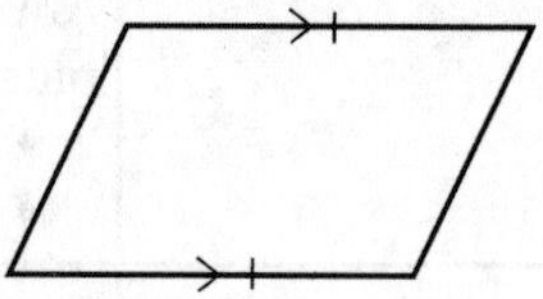
Two pairs of opposite angles are congruent.	
Two pairs of consecutive angles are supplementary.	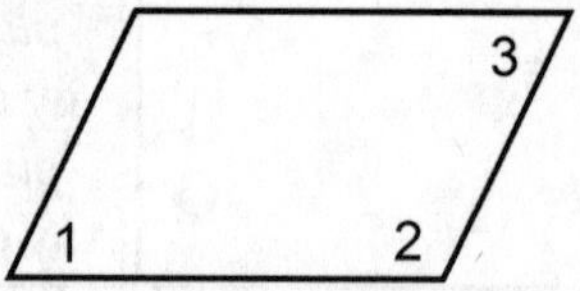 ∠1 and ∠2, ∠2 and ∠3 are supplementary
The diagonals bisect each other.	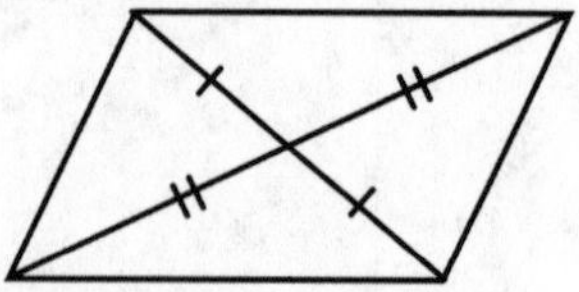

Practice Exercises

1 Parallelogram *PARK* has congruent and perpendicular diagonals. Which of the following describes *PARK*?

   (1) square

   (2) rectangle, but not rhombus

   (3) rhombus, but not rectangle

   (4) trapezoid, but not rectangle or rhombus

2 Quadrilateral *JUMP* has the following properties: $JU = MP$, $UM = PJ$, and $JU = UM$. How can *JUMP* be classified?

   (1) parallelogram only

   (2) parallelogram and rectangle

   (3) parallelogram and rhombus

   (4) not enough information to make any classification

3 In parallelogram *RSTU*, $m\angle R = 8x + 12$ and $m\angle T = 4x + 24$. What is $m\angle R$?

   (1) 144°   (3) 36°

   (2) 72°   (4) 18°

4 In parallelogram *ABCD*, the measures of $\angle B$ and $\angle C$ are in a 3:2 ratio. What is the measure of $\angle B$?

   (1) 120°   (3) 72°

   (2) 108°   (4) 60°

5 What are the measures of ∠1 and ∠2 in the given trapezoid?

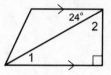

(1) m∠1 = 24° and m∠2 = 24°
(2) m∠1 = 66° and m∠2 = 24°
(3) m∠1 = 66° and m∠2 = 66°
(4) m∠1 = 24° and m∠2 = 66°

6 Given rhombus *ABCD* with diagonals $\overline{AC}$ and $\overline{BD}$ intersecting at *E*. If m∠*DAE* = 41°, what is the measure of ∠*CBE*?

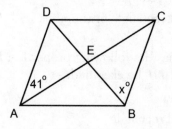

(1) 20.5°   (3) 45°
(2) 41°   (4) 49°

7 *ABCD* is a rhombus with diagonals $\overline{AC}$ and $\overline{BD}$ intersecting at *E*. If *AC* = 16 and *BD* = 12, what is the length of $\overline{AB}$?

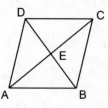

(1) 14   (3) 10
(2) $4\sqrt{14}$ (4) 28

8 Find the measure of ∠C in the parallelogram ABCD.

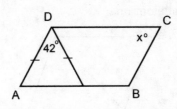

(1) 42°   (3) 69°

(2) 58°   (4) 71°

9 Diagonal $\overline{QS}$ is drawn in rectangle QRST. If m∠RQS = (4x + 2)° and m∠QSR = (3x + 11)°, what is m∠RQS?

(1) 11°   (3) 46°

(2) 44°   (4) 90°

10 In rectangle CARS, diagonals $\overline{SA}$ and $\overline{CR}$ intersect at X. If m∠RSX = 28°, what is the measure of ∠SXC?

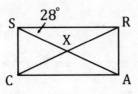

(1) 56°   (3) 66°

(2) 62°   (4) 68°

11 Square STAR has diagonals that intersect at P. If the perimeter of STAR is 24, what is the length of $\overline{RP}$ in simplest radical form?

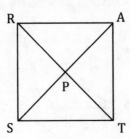

12 Given: $\angle E \cong \angle G$ and $\angle EDF \cong \angle GFD$

Prove: $DEFG$ is a parallelogram

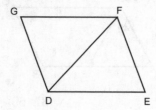

13 Given: $\overline{DKB}$, $\overline{DLC}$, $PLDK$ is a rhombus, and $\overline{BK} \cong \overline{CL}$

Prove: $\angle B \cong \angle C$

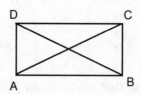

14 Given: Rectangle $ABCD$ with diagonals $\overline{AC}$ and $\overline{BD}$

Prove: The diagonals of a rectangle are congruent $\left(\overline{AC} \cong \overline{BD} \right)$

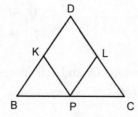

15 Given: $\angle BCA \cong \angle DAC$, $\angle BAC \cong \angle BCA$, $\overline{BC} \cong \overline{AD}$

Prove: $ABCD$ is a rhombus

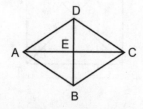

Solutions

1 A parallelogram with perpendicular diagonals is a rhombus, and a parallelogram with congruent diagonals is a rectangle. A figure with the properties of a rectangle and a rhombus is a square.

The correct choice is (**1**).

2 Opposites sides $\overline{JU}$ and $\overline{MP}$, and $\overline{UM}$ and $\overline{PJ}$ are congruent, making $JUMP$ a parallelogram. Consecutive sides $\overline{JU}$ and $\overline{UM}$ are congruent, making $JUMP$ a rhombus.

The correct choice is (**3**).

3     $m\angle R = m\angle T$ Opposite angles of a parallelogram are congruent

$$8x + 12 = 4x + 24$$
$$4x + 12 = 24$$
$$4x = 12$$
$$x = 3$$
$$m\angle R = 8(3) + 12 \quad \text{Substitute } x = 3$$
$$= 24 + 12$$
$$= 36°$$

The correct choice is (**3**).

4 Let $m\angle B = 3x$ and $m\angle C = 2x$, and apply the parallelogram property that consecutive angles are supplementary.

$$m\angle B + m\angle C = 180°$$
$$3x + 2x = 180$$
$$5x = 180$$
$$x = 36$$
$$m\angle B = 3(36) \quad \text{Substitute } x = 36$$
$$= 108°$$

The correct choice is (**2**).

5 The diagonal is a transversal forming a pair of congruent alternate interior angles, so m∠1 = 24°. The triangle angle sum theorem can be used on the bottom triangle to find ∠2.

$$m\angle 1 + m\angle 2 + 90° = 180°$$
$$24° + m\angle 2 + 90° = 180°$$
$$m\angle 2 + 114° = 180°$$
$$m\angle 2 = 66°$$

The correct choice is **(4)**.

6 m∠BCE = 41°   ∠DAE and ∠BEC are congruent alternate interior angles

m∠BEC = 90°   Diagonals of a rhombus are perpendicular

m∠CBE + m∠BEC + m∠BCE = 180°   Triangle angle sum theorem

$$m\angle CBE + 90° + 41° = 180°$$
$$m\angle CBE + 131° = 180°$$
$$m\angle CBE = 49°$$

The correct choice is **(4)**.

7 Use the rhombus property that the diagonals bisect each other and are perpendicular.

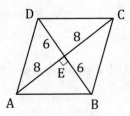

$$AE = \frac{1}{2}AC \quad \text{Diagonals of a rhombus bisect each other}$$

$$= \frac{1}{2}(16)$$

$$= 8$$

$$BE = \frac{1}{2}BD$$

$$= \frac{1}{2}(12)$$

$$= 6$$

$$\triangle AEB \text{ is a right triangle} \quad \text{Diagonals of a rhombus are perpendicular}$$

$$(AE)^2 + (BE)^2 = (AB)^2 \quad \text{Pythagorean theorem}$$

$$6^2 + 8^2 = (AB)^2$$

$$100 = (AB)^2$$

$$AB = 10$$

The correct choice is (**3**).

8 $m\angle A = \dfrac{180° - 42°}{2} \quad$ Isosceles triangle theorem

$$= 69°$$

$$m\angle C = m\angle A \quad \text{Opposite angles are congruent}$$

$$= 69°$$

The correct choice is (**3**).

9 Start by sketching the figure.

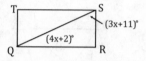

$\triangle QSR$ is a right triangle with a right angle at $\angle R$, so $\angle RQS$ and $\angle QSR$ must be complementary.

$$m\angle RQS + m\angle QSR = 90°$$
$$4x + 2 + 3x + 11 = 90$$
$$7x + 13 = 90$$
$$7x = 77$$
$$x = 11$$
$$m\angle RQS = 4(11) + 2 \quad \text{Substitute } x = 11$$
$$m\angle RQS = 46°$$

The correct choice is **(3)**.

10 $m\angle CSX + m\angle RSX = 90°$ Angles in a rectangle are right
                                    angles
$$m\angle CSX + 28° = 90°$$
$$m\angle CSX = 62°$$

The diagonals of a rectangle are congruent and bisect each other; therefore, $\overline{SX} \cong \overline{CX}$, making $\triangle SXC$ isosceles.

$m\angle SCX = m\angle CSX$ Base angles of an isosceles triangle
                                  are congruent
$$= 62°$$

$m\angle CSX + m\angle SCX + m\angle SXC = 180°$ Triangle angle sum
                                                    theorem
$$62° + 62° + m\angle SXC = 180°$$
$$124° + m\angle SXC = 180°$$
$$m\angle SXC = 56°$$

The correct choice is **(1)**.

11 $RS = \dfrac{1}{4}$ perimeter   All sides of a square are congruent

$\qquad = \dfrac{1}{4}(24)$

$\qquad = 6$

$\qquad\qquad RP = PS$ Diagonals of a square bisect each other and are congruent

$(RP)^2 + (PS)^2 = (RS)^2$ Pythagorean theorem

$(RP)^2 + (RP)^2 = 6^2$

$\qquad 2(RP)^2 = 36$

$\qquad\quad (RP)^2 = 18$

$\qquad\qquad RP = \sqrt{18}$

$\qquad\qquad\quad = 3\sqrt{2}$

12

Statement	Reason
1. $\angle EDF \cong \angle GFD$	1. Given
2. $\overline{ED} \,//\, \overline{FG}$	2. Two lines are parallel if the alternate interior angles formed are congruent
3. $\angle E \cong \angle G$	3. Given
4. $\overline{FD} \cong \overline{FD}$	4. Reflexive property
5. $\triangle EDF \cong \triangle GFD$	5. AAS
6. $\angle EFD \cong \angle GDF$	6. CPCTC
7. $\overline{GD} \,//\, \overline{FE}$	7. Two lines are parallel if the alternate interior angles formed are congruent
8. $DEFG$ is a parallelogram	8. A quadrilateral with two pairs of opposite parallel sides is a parallelogram

13

Statement	Reason
1. $\overline{DKB}$, $\overline{DLC}$, and *PLDK* is a rhombus	1. Given
2. $\overline{KD} \cong \overline{LD}$	2. Consecutive sides of a rhombus are congruent
3. $\overline{BK} \cong \overline{CL}$	3. Given
4. $\overline{BK} + \overline{KD} \cong \overline{CL} + \overline{LD}$	4. Addition postulate
5. $\overline{BD} \cong \overline{CD}$	5. Partition property
6. $\angle B \cong \angle C$	6. Angles opposite congruent sides in triangles are congruent

14 The strategy is to prove $\triangle ABC \cong \triangle BAD$ and then apply CPCTC.

Statement	Reason
1. Rectangle *ABCD* with diagonals $\overline{AC}$ and $\overline{BD}$	1. Given
2. $\angle ABC$ and $\angle BAD$ are right angles	2. All angles in a rectangle are right angles
3. $\angle ABC \cong \angle BAD$	3. Right angles are congruent
4. $\overline{AD} \cong \overline{BC}$	4. Opposite sides of a rectangle are congruent
5. $\overline{AB} \cong \overline{AB}$	5. Reflexive property
6. $\triangle ABC \cong \triangle BAD$	6. SAS
7. $\overline{AC} \cong \overline{BD}$	7. CPCTC

15 The strategy is to first prove $ABCD$ is a parallelogram and then prove $\triangle ABC$ is an isosceles triangle, with legs $\overline{AB} \cong \overline{BC}$.

Statement	Reason
1. $\angle BCA \cong \angle DAC$	1. Given
2. $\overline{BC} \parallel \overline{AD}$	2. Two lines are parallel if the alternate interior angles formed by a transversal are congruent
3. $\overline{BC} \cong \overline{AD}$	3. Given
4. $ABCD$ is a parallelogram	4. A quadrilateral is a parallelogram if one pair of sides is parallel and congruent
5. $\angle BAC \cong \angle BCA$	5. Given
6. $\triangle ABC$ is isosceles	6. A triangle with congruent base angles is isosceles
7. $\overline{AB} \cong \overline{BC}$	7. Sides opposite congruent angles in a triangle are congruent
8. $ABCD$ is a rhombus	8. A parallelogram with a pair of consecutive congruent sides is a rhombus

3.10 COORDINATE GEOMETRY PROOFS

Distance, midpoint, and slope are the tools used on the coordinate plane to prove properties of a figure. They are used as follows:

Quantity	Formula	Use
Slope	$m = \dfrac{y_2 - y_1}{x_2 - x_1}$ or $\dfrac{\text{rise}}{\text{run}}$	Prove segments or lines are parallel (slopes are equal) or perpendicular (slopes are negative reciprocals)
Midpoint	$x_{\text{MP}} = \dfrac{1}{2}(x_1 + x_2)$ $y_{\text{MP}} = \dfrac{1}{2}(y_1 + y_2)$	Prove segments bisect each other (midpoints are concurrent)
Length	$d = \sqrt{(x_2 - x_1)^2 + (y_2 - y_1)^2}$	Prove segments are congruent (distances between endpoints are equal)

The steps to writing a good coordinate geometry proof are

- *Graph the points*—The graph will help you plan a strategy, check your work, and help with some calculations.
- *Plan a strategy*—Determine what property you will demonstrate and what calculations are needed.
- *Perform the calculations*—Write down the general equations you are using and clearly label which segments correspond to which calculations.
- *Write a summary statement*—You must state in words a justification that explains why your calculations justify the proof.

PROVING PARALLELOGRAMS ON THE COORDINATE PLANE

A given quadrilateral can be proven to be a parallelogram by demonstrating one of the parallelogram properties. The three properties that can be easily proven are parallel sides, congruent sides, and diagonals bisecting each other. The calculations required and suggested summary statements are shown in the following table.

Property	Calculations	Summary Statement
2 pairs of opposite sides are parallel	Slope of each side (4 slope calculations)	The slopes of opposite sides are equal; therefore, both pairs of opposite sides are parallel. A quadrilateral with two pairs of opposite sides parallel is a parallelogram.
2 pairs of opposite sides are congruent	Length of each side (4 distance calculations)	The lengths of opposite sides are equal; therefore, both pairs of opposite sides are congruent. A quadrilateral with two pairs of opposite congruent sides is a parallelogram.
The diagonals bisect each other	Midpoints of the diagonals (2 midpoint calculations)	The diagonals have the same midpoint; therefore, they bisect each other. A quadrilateral whose diagonals bisect each other is a parallelogram.

PROVING SPECIAL PARALLELOGRAMS ON THE COORDINATE PLANE

To prove a quadrilateral is a special parallelogram, a good strategy is to first prove the figure is a parallelogram and then to demonstrate one of the special properties shown in the table. For squares, you need to show that the figure is both a rectangle and a rhombus.

First show the figure is a parallelogram, then...

Figure	Property	Calculation	Example of Summary Statement
Rhombus	Perpendicular diagonals	Slope of each diagonal	The slopes of the diagonals are negative reciprocals; therefore, they are ⊥. A parallelogram with ⊥ diagonals is a rhombus.
	Two consecutive congruent sides	Length of two consecutive sides	The lengths of 2 consecutive sides are equal, so they are ≅. A parallelogram with consecutive ≅ sides is a rhombus.
Rectangle	A right angle	Slope of two consecutive sides	The slopes of two consecutive sides are negative reciprocals, so the sides are ⊥ and form a right angle. A parallelogram with a right angle is a rectangle.
	Congruent diagonals	Length of each diagonal	The lengths of the diagonals are ≅, so they are congruent. A parallelogram with congruent diagonals is a rectangle.
Square	Show the figure is both a rectangle and a rhombus.		A figure that is both a rectangle and a rhombus is a square.

One strategy to aid in remembering which properties to prove is to work only with the diagonals:

- Midpoints for a parallelogram
- Midpoints and slope for a rhombus
- Midpoints and length for a rectangle

OTHER COORDINATE GEOMETRY PROOFS

A quadrilateral can be shown to be a trapezoid using slope calculations to show two sides are parallel.

Triangles can be classified using coordinate geometry calculations:

Isosceles triangle—distance calculation to show two sides are congruent
Equilateral—distance calculation to show three sides are congruent
Right triangle—slope calculation to show two sides are perpendicular

Coordinate geometry can be used to prove special segments in triangles. The following table summarizes some of the segments and points of concurrency, and the calculations needed.

Segment	Calculation Needed	Summary Statement
Median	Midpoint	The segment has endpoints at a vertex and midpoint, so it is a median.
Altitude	Two slopes	The slopes are negative reciprocals, so the segment is ⊥ to the opposite side, making it an altitude.
Perpendicular bisector	Slope and midpoint	The slopes are negative reciprocals, making the segment ⊥, and the segment passes through the midpoint, making it a bisector. Therefore, it is a perpendicular bisector.
Midsegment	Two midpoints	The segment's endpoints are the midpoints of sides of the triangle; therefore, it is a midsegment.
Circumcenter	Three lengths	The distance from the point to the three vertices are equal, so it is equidistant from each vertex. Therefore, the point is the circumcenter.
Centroid	Two lengths	The distances from the point to the vertices and the distance from the point to the opposite midpoint are in a 1:2 ratio; therefore, the point is a centroid.

Practice Exercises

1 The coordinates of quadrilateral *BGRM* are *B*(0, 2), *G*(6, 0), *R*(9, 9), and *M*(3, 11).

Prove *BGRM* is a rectangle.

2 △*GYP* has coordinates *G*(1, 1) *Y*(7, 5) and *P*(3, 11).

Prove △*GYP* is an isosceles right triangle.

3 Given parallelogram *FLIP* with vertices *F*(2, 9), *L*(*h*, *k*), *I*(11, 6), and *P*(1, 2), find the values of *h* and *k*.

4 Given quadrilateral *CUTE* with coordinates *C*(−5, −4), *U*(−3, 4), *T*(5, 6), and *E*(3, −2), prove *CUTE* is a rhombus.

5 Quadrilateral *PLOW* has coordinates *P*(1, −4), *L*(−1, 4), *O*(7, 6), and *W*(9, −2).

Prove *PLOW* is a square.

6 Quadrilateral *ABCD* has coordinates *A*(1, 2), *B*(9, 4), *C*(5, 8), and *D*(1, 7).

Prove *ABCD* is a trapezoid.

7 Segment $\overline{MG}$ has coordinates $M(0, 2)$ and $G(4, 0)$. Point T is the midpoint of $\overline{MG}$. $M''G''$ is the image of $\overline{MG}$ after a $90°$ rotation about T followed by a dilation centered about T with a scale factor of 2.

a) State the coordinates of M'' and G'' and graph $\overline{M''G''}$.

b) Prove $MG''GM''$ is a rhombus.

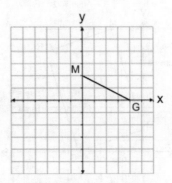

8 $\triangle ABC$ has coordinates $A(1, 5)$, $B(7, -1)$, and $C(13, 11)$. If point G has coordinates $(7, 5)$, prove G is the centroid of $\triangle ABC$.

9 $\triangle DOG$ has coordinates $D(4, 3)$, $O(9, 8)$, and $G(1, 12)$. If point A has coordinates $(3, 6)$, prove $\overline{OA}$ is an altitude of $\triangle DOG$.

10 Rectangle $ABCD$ has coordinates $A(4, 12)$, $B(13, 3)$, $C(9, -1)$, and $D(0, 8)$. Points F and G have coordinates $F(2, 10)$ and $G(7, 5)$, and $AHGF$ is also a rectangle. Is rectangle $ABCD$ similar to rectangle $AHGF$? Justify your answer.

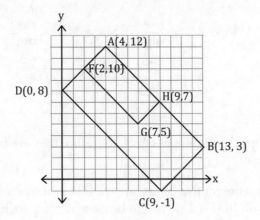

Solutions

1 There are several different approaches to proving a quadrilateral is a rectangle. Here, we will calculate the slopes of the 4 sides. This will let us show opposite sides have the same slope and are parallel, and consecutive sides have negative reciprocal slopes and are perpendicular. *BGRM* is graphed in the accompanying diagram.

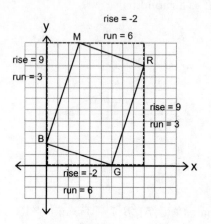

Using the grid, we can calculate slope from slope $= \dfrac{\text{rise}}{\text{run}}$. The rise and run for each side of *BGRM* are shown in the figure.

$$\text{Slope } \overline{BM} = \frac{\text{rise}}{\text{run}} \qquad \text{Slope } \overline{GR} = \frac{\text{rise}}{\text{run}}$$
$$= \frac{9}{3} \qquad\qquad\qquad = \frac{9}{3}$$
$$= 3 \qquad\qquad\qquad\quad = 3$$
$$\text{Slope } \overline{BG} = \frac{\text{rise}}{\text{run}} \qquad \text{Slope } \overline{MR} = \frac{\text{rise}}{\text{run}}$$
$$= \frac{-2}{6} \qquad\qquad\qquad = \frac{-2}{6}$$
$$= -\frac{1}{3} \qquad\qquad\quad = -\frac{1}{3}$$

The slopes of opposite sides are equal, therefore they are parallel and *BRGM* is a parallelogram.

The slopes of consecutive sides are negative reciprocals, making them perpendicular. *BGRM* is a rectangle because it is a parallelogram with right angles.

2 The strategy is to show that $\overline{GY} \cong \overline{YP}$ using distance calculations and $\overline{GY} \perp \overline{YP}$ using slope calculations.

$$\text{slope} = \frac{y_2 - y_1}{x_2 - x_1}$$

$$\text{slope } \overline{GY} = \frac{5 - 1}{7 - 1} \qquad\qquad \text{slope } \overline{YP} = \frac{11 - 5}{3 - 7}$$

$$= \frac{4}{6} = \frac{2}{3} \qquad\qquad\qquad\qquad = -\frac{6}{4} = -\frac{3}{2}$$

$$\text{length} = \sqrt{\left(x_1 - x_2\right)^2 + \left(y_1 - y_2\right)^2}$$

$$\text{length } \overline{GY} = \sqrt{\left(7 - 1\right)^2 + \left(5 - 1\right)^2}$$

$$= \sqrt{36 + 16}$$

$$= \sqrt{52}$$

$$\text{length } \overline{YP} = \sqrt{\left(7 - 3\right)^2 + \left(5 - 11\right)^2}$$

$$= \sqrt{16 + 36}$$

$$= \sqrt{52}$$

$\overline{GY}$ is perpendicular to $\overline{YP}$ because their slopes are negative reciprocals, and $\overline{GY} \cong \overline{YP}$ because the lengths GY and YP are equal. Therefore, $\triangle GYP$ is a right isosceles triangle.

3 The opposite sides of a parallelogram are parallel and congruent. The strategy is to locate point L so that $FL = PI$ and the slopes of $\overline{FL}$ and $\overline{PI}$ are equal.

$$\text{slope} = \frac{y_2 - y_1}{x_2 - x_1}$$

$$\text{slope } \overline{PI} = \frac{6 - 2}{11 - 1}$$

$$= \frac{4}{10}$$

$$\text{slope } \overline{FL} = \frac{4}{10} \qquad \text{Parallel lines have equal slopes.}$$

Since slope is also equal to $\frac{\text{rise}}{\text{run}}$, locate point L by adding the rise to the y-coordinate of F, and add the run to the x-coordinate of x.

$$x\text{-corrdinate of } L = 2 + 10$$
$$= 12$$
$$y\text{-coordinate of } L = 9 + 4$$
$$= 13$$

The coordinates of L are (12, 13).

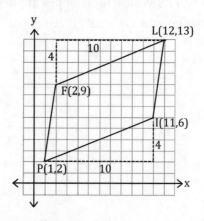

4 One strategy is to first prove the figure is a parallelogram by showing the diagonals bisect each other using midpoint calculations. Then prove it is a rhombus by showing the diagonals are perpendicular using slope calculations.

$$\text{midpoint} = \frac{x_1 + x_2}{2}, \frac{y_1 + y_2}{2}$$

midpoint $\overline{CT}$: $\frac{-5+5}{2}, \frac{-4+6}{2}$ midpoint $\overline{EU}$: $\frac{3+(-3)}{2}, \frac{-2+4}{2}$

$(0,1)$ $(0,1)$

$$\text{slope} = \frac{y_2 - y_1}{x_2 - x_1}$$

$$\text{slope } \overline{CT} = \frac{6-(-4)}{5-(-5)} \qquad \text{slope } \overline{EU} = \frac{4-(-2)}{-3-3}$$

$$= \frac{10}{10} \qquad\qquad\qquad = -\frac{6}{6}$$

$$= 1 \qquad\qquad\qquad\quad = -1$$

The diagonals of $CUTE$ bisect each because they have the same midpoint, making it a parallelogram. The diagonals are also perpendicular because their slopes are negative reciprocals. Therefore, $CUTE$ is a rhombus.

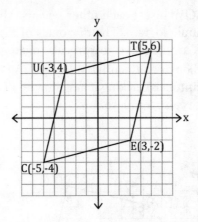

5 First prove the figure is a parallelogram by showing the diagonals bisect each other using midpoint calculations; then show it is a rhombus with perpendicular diagonals using slope. Finally, show it is a rectangle with a pair of perpendicular sides using slope.

$$\text{midpoint} = \frac{x_1 + x_2}{2}, \frac{y_1 + y_2}{2}$$

midpoint $\overline{PO}$: $\frac{1+7}{2}, \frac{-4+6}{2}$ 　　 midpoint $\overline{WL}$: $\frac{9+(-1)}{2}, \frac{-2+4}{2}$

$(4, 1)$ 　　　　　　　　　　 $(4, 1)$

$$\text{slope} = \frac{y_2 - y_1}{x_2 - x_1}$$

$$\text{slope } \overline{PO} = \frac{6-(-4)}{7-1} \qquad \text{slope } \overline{WL} = \frac{4-(-2)}{-1-9}$$

$$= \frac{10}{6} \qquad\qquad\qquad = -\frac{6}{10}$$

$$\text{slope } \overline{PW} = \frac{-2-(-4)}{9-1} \qquad \text{slope } \overline{LP} = \frac{-4-4}{1-(-1)}$$

$$= \frac{2}{8} \qquad\qquad\qquad = -\frac{8}{2}$$

The diagonals of *PLOW* bisect each other because they have the same midpoint, making it a parallelogram. The diagonals of *PLOW* are perpendicular because their slopes are negative reciprocals, making it a rhombus. *PLOW* has a pair of perpendicular consecutive sides because $\overline{PW}$ and $\overline{LP}$ have slopes that are negative reciprocals, making it a rectangle. *PLOW* is both a rectangle and a rhombus, so it is also a square.

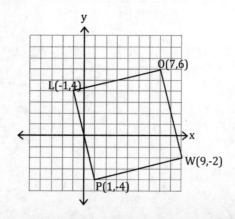

6 A good time-saving strategy is to graph the figure first to determine the pair of sides that are likely to be parallel. The parallel sides appear to be $\overline{AB}$ and $\overline{CD}$, and the legs are $\overline{AD}$ and $\overline{BC}$.

$$\text{slope} = \frac{y_2 - y_1}{x_2 - x_1}$$

$$\text{slope } \overline{AB} = \frac{4 - 2}{9 - 1} \qquad \text{slope } \overline{CD} = \frac{8 - 7}{5 - 1}$$

$$= \frac{2}{8} \qquad\qquad\qquad = \frac{1}{4}$$

$$= \frac{1}{4}$$

$\overline{AB}$ is parallel to $\overline{CD}$ because their slopes are equal, making $ABCD$ a trapezoid.

7 a) First find the coordinates of the midpoint T using the midpoint formula; then use T to graphically find the coordinates of M' and G'.

$$x_{MP} = \frac{1}{2}(x_1 + x_2)$$

$$= \frac{1}{2}(0 + 4)$$

$$= 2$$

$$y_{MP} = \frac{1}{2}(y_1 + y_2)$$

$$= \frac{1}{2}(2 + 0)$$

$$= 1$$

The coordinates of T are $(2, 1)$.

Rotating a segment 90° about its midpoint results in a segment congruent and perpendicular to the pre-image.

$$\text{slope of } \overline{MG} = \frac{y_2 - y_1}{x_2 - x_1}$$

$$= \frac{0 - 2}{4 - 0}$$

$$= -\frac{1}{2}$$

Let $\overline{M'G'}$ be the image of $\overline{MG}$ after the 90° rotation. The slope of $\overline{M'G'}$ is the negative reciprocal, or 2. The coordinates of M' and G' can be found graphically by starting at T and moving with a slope of 2 in either direction for a length equal to TM and TG. From the graph, the coordinates are $M'(1, -1)$ and $G'(3, 3)$.

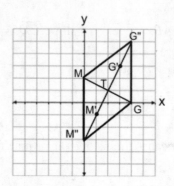

A dilation through midpoint T will extend the segment the same distance on either side of the midpoint because T remains the midpoint. Graph $\overline{M''G''}$ collinear with $\overline{M'G'}$, but twice as long.

b) $MG''GM''$ must be a rhombus. T is the midpoint of both of its diagonals. The diagonals bisect each other making it a parallelogram. The 90° rotation resulted in $\overline{MG} \perp \overline{M''G''}$. A parallelogram with perpendicular diagonals is a rhombus.

8 The centroid is the point of concurrency of the medians of a triangle, so one strategy is to graph two medians and look for the point of intersection. A median is the segment from a vertex to the opposite midpoint, so the first step is to find the midpoints of two sides.

$$\text{midpoint} = \frac{x_1 + x_2}{2}, \frac{y_1 + y_2}{2}$$

let D be the of $\overline{AB}$: $\frac{1+7}{2}, \frac{5+(-1)}{2}$

$D(4,2)$

let E be the of $\overline{AC}$: $\frac{1+13}{2}, \frac{5+11}{2}$

$E(7,8)$

Graphing $\overline{CD}$ and $\overline{BE}$, the point of intersection is $G(7,5)$. Therefore, G is the centroid of $\triangle ABC$.

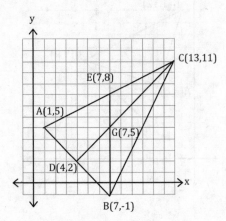

Algebraically, you could have also used the distance formula to calculate DG and GC and then show $DG = \frac{1}{2}GC$.

9 An altitude is a segment from a vertex perpendicular to the opposite side of a triangle. $\overline{AO}$ is an altitude if it is perpendicular to $\overline{GD}$.

$$\text{slope} = \frac{y_2 - y_1}{x_2 - x_1}$$

$$\text{slope } \overline{AO} = \frac{8 - 6}{9 - 3} \qquad \text{slope } \overline{GD} = \frac{12 - 3}{1 - 4}$$

$$= \frac{2}{6} \qquad\qquad\qquad = -\frac{9}{3}$$

$$= \frac{1}{3} \qquad\qquad\qquad = -\frac{3}{1}$$

$\overline{AO}$ is perpendicular to $\overline{GD}$ because their slopes are negative reciprocals, so $\overline{AO}$ is an altitude of $\triangle DOG$.

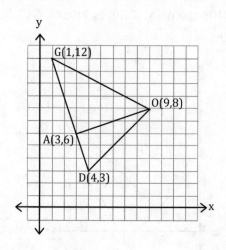

10 Since both figures are rectangles, all pairs of corresponding angles must be congruent. The lengths of two pairs of corresponding sides must be calculated to see if they are in the same ratio. The other two pairs must be in the same ratio since opposite sides of a rectangle are congruent.

The corresponding lengths to be checked are AB and AH, and AD and AF.

$$\text{length} = \sqrt{(x_1 - x_2)^2 + (y_1 - y_2)^2}$$

$$\text{length } AB = \sqrt{(4 - 13)^2 + (12 - 3)^2}$$

$$= \sqrt{81 + 81}$$

$$= \sqrt{162}$$

$$\text{length } AH = \sqrt{(4 - 9)^2 + (12 - 7)^2}$$

$$= \sqrt{25 + 25}$$

$$= \sqrt{50}$$

$$\text{length } AD = \sqrt{(4 - 0)^2 + (12 - 8)^2}$$

$$= \sqrt{16 + 16}$$

$$= \sqrt{32}$$

$$\text{length } AF = \sqrt{(4 - 2)^2 + (12 - 10)^2}$$

$$= \sqrt{4 + 4}$$

$$= \sqrt{8}$$

Calculate the ratios of the two pairs of corresponding sides.

$$\frac{AB}{AH} = \frac{\sqrt{162}}{\sqrt{50}} \qquad \frac{AD}{AF} = \frac{\sqrt{32}}{\sqrt{8}}$$

$$= \sqrt{\frac{81}{25}} \qquad \qquad = \sqrt{4}$$

$$= \frac{9}{5} \qquad \qquad \qquad = 2$$

The rectangles are not similar because the corresponding sides are not in the same ratio.

3.11 CIRCLES

DEFINITIONS AND BASIC THEOREMS

Segment and Line Definitions in Circles

Segment or Line	Definition
Radius	A segment with one endpoint at the center of the circle and one endpoint on the circle
Chord	A segment with both endpoints on the circle
Diameter	A chord that passes through the center of the circle
Secant	A line that intersects a circle at exactly two points
Tangent	A line that intersects a circle at exactly one point
Point of tangency	The point at which a tangent intersects a circle

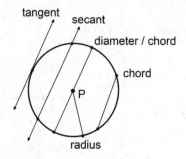

Some basic theorems of circles:

- All radii of a given circle are congruent.
- All circles are similar.
- Two circles are congruent if and only if their radii are congruent.

Since all circles are similar, any circle can be mapped to another using a similarity transformation comprised of a translation to move one center onto the other followed by a dilation to make the radii congruent.

An arc is a section of a circle. An arc may be major, minor, or a semicircle:

- *Minor arc*—an arc spanning less than 180°
- *Semicircle*—an arc spanning exactly 180°
- *Major arc*—an arc spanning more than 180°

When naming a semicircle or major arc, we add an additional point to show which arc is being specified. The accompanying figure shows a minor arc $\overset{\frown}{AB}$, semicircle $\overset{\frown}{ABC}$, and major arc $\overset{\frown}{ABD}$.

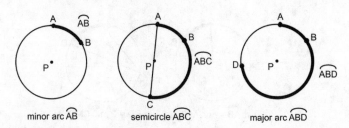

Some useful arc theorems:

- A diameter intercepts a semicircle.
- The sum of the arc measures around a circle equals 360°.
- The sum of the arc measures around a semicircle equals 180°.

CENTRAL AND INSCRIBED ANGLES

Central angle—an angle whose vertex is the center of a circle and whose rays intersect the circle

---

CENTRAL ANGLE THEOREM

The angle measure of an arc equals the measure of the central angle that intercepts the arc.

Inscribed angle—an angle whose vertex is on a circle and whose rays intersect the circle

---

INSCRIBED ANGLE THEOREM

The angle measure of an arc equals twice the measure of the inscribed angle that intercepts the arc.

The accompanying figure shows central angle $\angle APB$ and inscribed angle $\angle CMD$. Each angle cuts off, or intercepts, an arc on the circle. Central $\angle APB$ intercepts $\overset{\frown}{AB}$ and inscribed angle $\angle CMD$ intercepts $\overset{\frown}{CD}$.

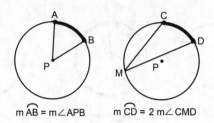

$$m \overset{\frown}{AB} = m\angle APB \qquad m \overset{\frown}{CD} = 2\, m\angle CMD$$

- Since the inscribed angle measures half the intercepted arc, any inscribed angle that intercepts a semicircle is a right angle.

CONGRUENT, PARALLEL, AND PERPENDICULAR CHORDS

CONGRUENT CHORD THEOREM

Congruent chords intercept congruent arcs on a circle.

Congruent arcs on a circle are intercepted by congruent chords.

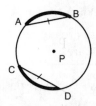

$$\overline{AB} \cong \overline{CD}, \quad \overset{\frown}{AB} \cong \overset{\frown}{CD}$$

PARALLEL CHORD THEOREM

The two arcs formed *between* a pair of parallel chords are congruent.

   If the two arcs formed *between* a pair of chords are congruent, then the chords are parallel.

$$\overline{AB} \parallel \overline{CD}, \quad \overset{\frown}{AC} \cong \overset{\frown}{BD}$$

CHORD–PERPENDICULAR BISECTOR THEOREM

The perpendicular bisector of any chord passes through the center of the circle.

   A diameter or radius that is perpendicular to a chord bisects the chord.

   A diameter or radius that bisects a chord is perpendicular to the chord.

In the accompanying figure, $\overline{AC}$ is perpendicular to diameter $\overline{BD}$ at E; therefore, E is the midpoint of $\overline{AC}$.

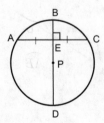

You can think of this theorem linking three properties of a diameter:

- Bisects another chord
- Is perpendicular to another chord
- Is a diameter

If any two are true, then the third must also be true.

TANGENT RADIUS THEOREM

- A diameter or radius to a point of tangency is perpendicular to the tangent.
- A line perpendicular to a tangent at the point of tangency passes through the center of the circle.

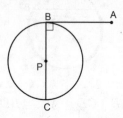

CONGRUENT TANGENT THEOREM

Given a circle and external point Q, segments between the external point and the two points of tangency are congruent.

In the accompanying figure, tangents $\overrightarrow{QA}$ and $\overrightarrow{QB}$ are both constructed from point Q, and $\overline{QA} \cong \overline{QB}$.

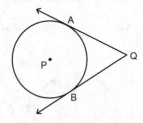

CYCLIC QUADRILATERALS

Quadrilaterals that are inscribed in a circle are called cyclic quadrilaterals.

CYCLIC QUADRILATERAL THEOREM

Opposite angles of cyclic quadrilaterals are supplementary.

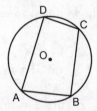

∠A and ∠C are supplementary.
∠B and ∠D are supplementary.

This property is proven by considering the intercepted arcs. Together, arcs $\overarc{ADC}$ and $\overarc{ABC}$ comprise the entire circle, so m$\overarc{ADC}$ + m$\overarc{ABC}$ = 360°. Since the intercepted angles measure $\frac{1}{2}$ the arc measures, m∠ABC + m∠ADC = 180°, and the angles are supplementary. The same justification can be used to show ∠BAD and ∠BCD are supplementary.

ANGLES FORMED BY INTERSECTING CHORDS, SECANTS, AND TANGENTS

Angles formed by intersecting chords, secants, and tangents follow three different relationships, depending on whether the vertex of the angle lies within, on, or outside the circle.

Vertex inside the circle	angle measure $= \frac{1}{2}$ the *sum* of the intercepted arcs $$m\angle 1 = m\angle 2 = \frac{1}{2}\left(m\widehat{AC} + m\widehat{DB}\right)$$ $$m\angle 3 = m\angle 4 = \frac{1}{2}\left(m\widehat{BC} + m\widehat{AD}\right)$$
Vertex on the circle 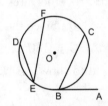	angle measure $= \frac{1}{2}$ of the intercepted arcs $$m\angle ABC = \frac{1}{2}m\widehat{BC}$$ $$m\angle DEF = \frac{1}{2}m\widehat{DF}$$
Vertex outside the circle 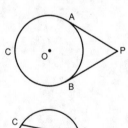 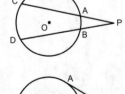 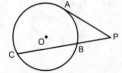	angle measure $= \frac{1}{2}$ the *difference* of the intercepted arcs $$m\angle P = \frac{1}{2}\left(m\widehat{ACB} - m\widehat{AB}\right)$$ $$m\angle P = \frac{1}{2}\left(m\widehat{CD} - m\widehat{AB}\right)$$ $$m\angle P = \frac{1}{2}\left(m\widehat{AC} - m\widehat{AB}\right)$$

SEGMENT RELATIONSHIPS IN INTERSECTING CHORDS, TANGENTS, AND SECANTS

The relationships between segment lengths formed by intersecting chords, tangents, and secants are shown in the following table.

Intersecting chords	products of the parts are equal $$a \cdot b = c \cdot d$$
Intersecting secants	outside · whole = outside · whole $$PA \cdot PC = PB \cdot PD$$
Intersecting tangent and secant	outside · whole = outside · whole $$PA^2 = PB \cdot PC$$

RADIAN MEASURE, ARC LENGTH, AND SECTORS

The **radian** is a unit of angle measure. 2π radians is equal to one complete revolution around a circle, or $360°$. Convert between units using:

$$\text{radians} = \frac{\pi}{180°} \cdot \text{degrees}$$

$$\text{degrees} = \frac{180°}{\pi} \cdot \text{radians}$$

- Besides having an angle measure, an arc has a length. The arc and its central angle also enclose a region called a sector.

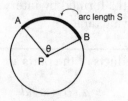

Arc with length S
intercepted by angle θ

Sector APB intercepted
by an angle θ

Arc length and sector area can be calculated from the measure of the central angle and the radius of the circle using the following formulas:

Unit of Angle Measure	Arc Length	Sector Area
Degrees	$\dfrac{\pi}{180°}R \cdot \theta$	$\dfrac{1}{360°}\pi R^2 \cdot \theta$
Radians	$R \cdot \theta$	$\dfrac{1}{2}R^2 \cdot \theta$

θ is the measure of the central angle.

Practice Exercises

1 Which of the following is a precise definition of a circle?

(1) the set of points a fixed distance away from a given point

(2) the set of points equidistant from the two endpoints of a given diameter

(3) the set of points equidistant from a given center point and a given line

(4) a closed figure with no angles

2 △STY is inscribed in circle P. Which of the following is *not necessarily* true?

(1) Point P is equidistant from S, T, and Y.

(2) The perpendicular bisector of $\overline{ST}$ passes through P.

(3) The measure of angle Y equals one-half the measure of $\overset{\frown}{ST}$.

(4) A tangent to the circle at S forms a right angle with $\overline{ST}$.

3 △FLY is inscribed in circle O. If FL = LY and m$\overset{\frown}{FY}$ = 80°, what is the measure of ∠F?

(1) 10°   (3) 70°

(2) 40°   (4) 80°

4 In ⊙P, ∠APB is a central angle and ∠ADB is an inscribed angle. If m∠APB = 110°, find m∠ADB.

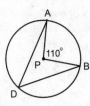

(1) 55°   (3) 70°

(2) 60°   (4) 80°

5 The clock in the tower of the Ridgefield town hall is in the shape of a circle 15 feet in diameter. What is the length of the arc around the clock whose central angle is defined by the hour and minute hands at 4:00 P.M.?

(1) 15.7 ft   (3) 31.4 ft

(2) 18.6 ft   (4) 60 ft

6 In circle Z, arc $\overset{\frown}{KP}$ is intercepted by central angle $\angle KZP$. If $m\angle KZP = 1.2$ radians and $ZK = 7$ cm, what is the length of $\overset{\frown}{KP}$?

(1) 5.8 cm   (3) 10.1 cm

(2) 8.4 cm   (4) 26.4 cm

7 Circle O has a radius of 14 inches. If $\angle HOT$ is a central angle in circle O that measures $132°$, what is the area of sector HOT?

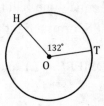

(1) 4.7 in^2 (3) 226 in^2

(2) 10 in^2 (4) 294 in^2

8 Given ⊙*P* with radius $\overline{PB}$, tangent $\overline{AB}$, and m∠*P* = 53°, what is the measure of ∠*A*?

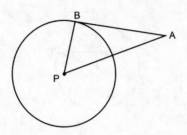

(1) 26.5°   (3) 53°

(2) 37°   (4) 60°

9 In ⊙*P*, chords $\overline{AB}$ and $\overline{CD}$ are parallel. If m$\overarc{AB}$ = 118° and m$\overarc{CD}$ = 78°, find m$\overarc{AC}$.

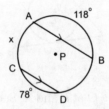

10 In ⊙*P*, chords $\overline{AB}$ and $\overline{CD}$ are congruent, m$\overarc{BC}$ = 60°, and m$\overarc{CD}$ = 80°. Find m∠*ABD*.

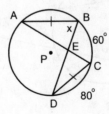

11 $\overline{BDP}$ is a radius of $\odot P$ and is perpendicular to $\overline{AC}$. If $AC = 24$ and $PC = 13$, find PD and DB.

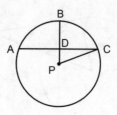

12 $\triangle RST$ is circumscribed about circle P. If $SX = 6$, $XR = 7$, and $TZ = 8$, what is the perimeter of $\triangle RST$?

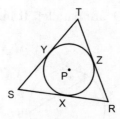

13 Secant $\overline{BEA}$ intersects circle O at E and A. Secant $\overline{BDR}$ intersects circle O at points D and R. If $m\widehat{AR} = 78°$ and $m\widehat{ED} = 35°$, what is the measure of $\angle B$?

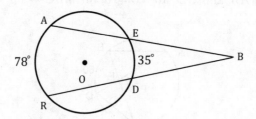

14 Secant $\overline{FGH}$ intersects circle O at G and H. Secant $\overline{FPQ}$ intersects circle O at P and Q. If $FG = 6$, $GH = 10$, and $FP = 8$, find the length of PQ.

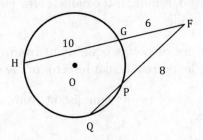

15 Tangents $\overline{TR}$ and $\overline{TQ}$ are drawn from point T to circle A. $\overline{RAS}$ is a diameter of circle A and m$\angle T = 46°$, what is the measure of $\overset{\frown}{SQ}$?

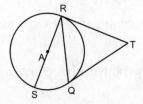

16 In $\odot O$, chords $\overline{HLI}$ and $\overline{KLJ}$ intersect at L. Prove: $HL \cdot LI = KL \cdot LJ$

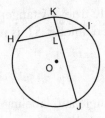

Solutions

1 Examine each choice:

(1) Correct—The fixed distance is the radius of the circle, and the given point is the center of the circle.

(2) Incorrect—The set of points equidistant from the two endpoints of a diameter would be the perpendicular bisector of the diameter.

(3) Incorrect—The set of points equidistant from a center point and line would be a parabola.

(4) Incorrect—Circles are closed figures with no angles, but not all closed figures with no angles are circles. A counterexample would be an ellipse.

The correct choice is (**1**).

2 Choice (1) is true—The center of the circumscribed circle is equidistant from the vertices of the triangle.

Choice (2) is true—A perpendicular bisector of a chord always passes through the center of the circle.

Choice (3) is true—The measure of the inscribed angle equals half the measure of the intercepted arc.

Choice (4) is not necessarily true—A tangent will make a right angle with a chord only if the chord is a diameter. $\overline{ST}$ is not necessarily a diameter.

The correct choice is (**4**).

3 Sketching the isosceles triangle and labeling the known arc, we see $\angle L$ is the inscribed angle that intercepts the 80° arc.

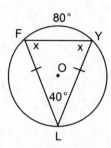

An inscribed angle measures half the intercepted arc:

$$m\angle L = \frac{1}{2}m\widehat{FY}$$

$$= \frac{1}{2}(80°)$$

$$= 40°$$

Since $FL = LY$, $m\angle F = m\angle Y$ from the isosceles triangle theorem, and we can use the angle sum theorem to find the measures.

$$m\angle F + m\angle L + m\angle Y = 180°$$

$$2x + 40° = 180°$$

$$2x = 140°$$

$$x = 70°$$

$$m\angle F = 70°$$

The correct choice is **(3)**.

4 The strategy is to use the central angle to find $m\widehat{AB}$, and then use the arc to find the inscribed angle.

$m\widehat{AB} = m\angle APB = 110°$ The central angle and arc have equal measure.

$m\angle ADB = \frac{1}{2}m\widehat{AB} = 55°$ The inscribed angle measure is half the intercepted arc.

The correct choice is **(1)**.

5 The numbers on the face of the clock divide the circle into 12 congruent central angles. Each central angle measures $\dfrac{360°}{12}$, or 30°. Therefore, the angle formed by the two hands on the clock measures $4 \cdot 30°$, or 120°. The radius $= \dfrac{1}{2}(15)$, or 7.5 ft. Using the formula for arc length,

$$S = 2\pi R \dfrac{\theta}{360°}$$
$$= 2\pi(7.5\,\text{ft})\dfrac{120°}{360°}$$
$$= 15.7\,\text{ft}$$

The correct choice is **(1)**.

6 Arc length, S, is given by the equation $S = R\theta$, where θ is the measure of the central angle in radians.

$$S = R\theta$$
$$= 7\,\text{cm} \cdot 1.2$$
$$= 8.4\,\text{cm}$$

The correct choice is **(2)**.

7 The area of a sector is given by $A = \dfrac{\theta}{360°}\pi R^2$, where θ is the central angle measured in degrees.

$$A = \dfrac{132°}{360°}\pi \cdot (14\,\text{in})^2$$
$$= 225.775\,\text{in}^2$$
$$= 226\,\text{in}^2$$

The correct choice is **(3)**.

8                $m\angle B = 90°$ radius perpendicular to a tangent
                                 at the point of tangency

$m\angle P + m\angle B + m\angle A = 180°$ triangle angle sum theorem

$53° + 90° + m\angle A = 180°$

$143° + m\angle A = 180°$

$m\angle A = 37°$

The correct choice is (**2**).

9 Arcs $\overset{\frown}{AC}$ and $\overset{\frown}{DB}$ are the congruent arcs between a pair of parallel chords.

$m\overset{\frown}{AB} + m\overset{\frown}{BD} + m\overset{\frown}{CD} + m\overset{\frown}{AC} = 360°$ circle–arc sum theorem

$118° + x + 78° + x = 360°$ parallel chord theorem

$196° + 2x = 360°$

$2x = 164°$

$x = 82°$

$m\overset{\frown}{AC} = 82°$

10 $\angle ABD$ is an inscribed angle, so its measure will be half of $m\overset{\frown}{AD}$.

$m\overset{\frown}{AB} + m\overset{\frown}{BC} + m\overset{\frown}{CD} + m\overset{\frown}{AD} = 360°$ circle–arc sum theorem

$m\overset{\frown}{AB} = m\overset{\frown}{CD} = 80°$ congruent chord theorem

$80° + 60° + 80° + m\overset{\frown}{AD} = 360°$

$220° + m\overset{\frown}{AD} = 360°$

$m\overset{\frown}{AD} = 140°$

$m\angle ABD = \frac{1}{2} m\overset{\frown}{AD}$

$= 70°$

11 Radius $\overline{BDP}$ is perpendicular to $\overline{AC}$, so it must also bisect $\overline{AC}$. The strategy is to calculate DC and then apply the Pythagorean theorem in $\triangle PDC$.

$$DC = \frac{1}{2}AC$$

$$DC = \frac{1}{2}(24)$$

$$DC = 12$$

$$(PD)^2 + (DC)^2 = PC^2 \qquad \text{Pythagorean theorem}$$

$$(PD)^2 + 12^2 = 13^2$$

$$(PD)^2 + 144 = 169$$

$$(PD)^2 = 25$$

$$PD = 5$$

$$PB = PC = 13 \quad \text{all radii of a circle are congruent}$$

$$PD + DB = PB \qquad \text{partition}$$

$$5 + DB = 13$$

$$DB = 8$$

12 $SY = SX$, $TY = TZ$, and $ZR = XR$ because each pair are tangents from the same point.

The perimeter is, therefore, $2 \cdot 6 + 2 \cdot 7 + 2 \cdot 8 = 42$.

13 The angle formed by two secants is equal to half the difference of the intercepted arcs.

$$m\angle B = \frac{1}{2}\left(m\widehat{AB} - m\widehat{ED}\right)$$

$$= \frac{1}{2}\left(78° - 35°\right)$$

$$= 21.5°$$

14 When two tangents from the same point intersect a circle, the segments formed follow the relationship:

$$\text{outside} \cdot \text{whole} = \text{outside} \cdot \text{whole}$$
$$FG \cdot FH = FP \cdot FQ$$
$$6 \cdot (6 + 10) = 8 \cdot (8 + PQ)$$
$$6 \cdot 16 = 8(8 + PQ)$$
$$96 = 64 + 8PQ$$
$$32 = 8PQ$$
$$PQ = 4$$

15 This problem requires applying multiple circle concepts. The strategy is to first use the fact $\triangle TRQ$ is isosceles to find m$\angle TRQ$ and m$\angle SRQ$ and then to apply the relationship that the measure of $\overarc{SQ}$ is twice the inscribed angle that intercepts it.

$$\overline{TR} \cong \overline{TQ} \qquad \qquad \text{congruent tangent theorem}$$

$$\text{m}\angle TRQ = \frac{1}{2}\left(180° - \text{m}\angle T\right) \quad \text{isosceles triangle theorem in } \triangle TRQ$$

$$\text{m}\angle TRQ = \frac{1}{2}\left(180° - 46°\right)$$

$$= 67°$$

$$\text{m}\angle SRT = 90° \qquad \qquad \text{tangent } RT \perp \text{diameter } \overline{RAS}$$

$$\text{m}\angle SRQ + \text{m}\angle TRQ = 90° \qquad \angle SRQ \text{ and } \angle TRQ \text{ are complementary}$$

$$\text{m}\angle SRQ + 67° = 90°$$

$$\text{m}\angle SRQ = 23°$$

$$\text{m}\overarc{SQ} = 2\text{m}\angle SRQ \qquad \text{inscribed angle theorem}$$

$$\text{m}\overarc{SQ} = 2\left(23°\right)$$

$$\text{m}\overarc{SQ} = 46°$$

16 The strategy is to sketch $\overline{HJ}$ and $\overline{KI}$, show $\triangle HLJ \sim \triangle KLI$, and then write a proportion between corresponding sides.

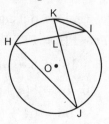

Statements	Reasons
1. Chords $\overline{HLI}$ and $\overline{KLJ}$ intersect at L	1. Given
2. Construct $\overline{IK}$ and $\overline{HJ}$	2. Two points define a segment
3. $\angle H \cong \angle K$ $\angle J \cong \angle I$	3. Inscribed angles that intercept the same arc are congruent
4. $\triangle HLJ \sim \triangle KLI$	4. AA
5. $\dfrac{HL}{KL} = \dfrac{LJ}{LI}$	5. Corresponding parts of similar triangles are proportional
6. $HL \cdot LI = KL \cdot LJ$	6. Cross products of a proportion are equal

3.12 SOLIDS

DEFINITIONS

- A *solid* is any 3-D figure that is fully enclosed.
- A *face* of a solid is any of the surfaces that bound the solid.
- A *polyhedron* is any solid whose faces are polygons.
- An *edge* is the intersection of two faces in a polyhedron.

PRISMS

A *prism* is a polyhedron with two congruent, parallel polygons for bases. The bases of a prism can have any shape. The accompanying figure shows 6 different prisms.

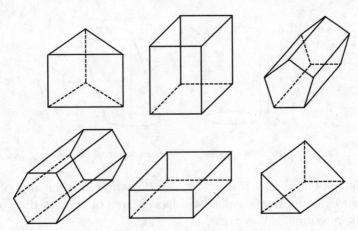

When working with prisms, keep in mind the following facts:

- The height of a prism, h, is the distance between the two bases shown in the accompanying figure.
- The lateral faces are all the faces other than the two parallel bases.
- A right prism has lateral edges that are perpendicular to the bases and lateral faces that are rectangles.
- The lateral edges of an oblique prism are not perpendicular to the bases, and the lateral faces are parallelograms.
- The volume of any prism is found with the formula $V = Bh$, where B is the area of the base.

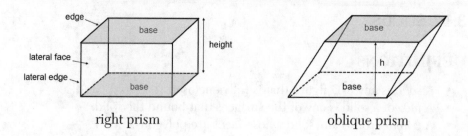

right prism   oblique prism

CYLINDERS

A *circular cylinder* is a solid figure with two parallel and congruent circular bases and a curved lateral area. The height is the perpendicular distance between the bases. As with the prisms, cylinders can be right or oblique as shown in the accompanying figure.

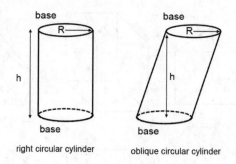

right circular cylinder   oblique circular cylinder

The volume of a cylinder is given by $V = Bh$, where B is the area of the base. In a circular cylinder, the circular base has an area of πR^2, so the volume formula can be rewritten as $V = \pi R^2 h$.

CONES AND PYRAMIDS

A *circular cone* is a solid with one circular base that comes to a point at an apex. A *pyramid* is a polyhedron having one polygonal base and triangles for lateral faces. The base of a pyramid can be any polygon, and the lateral faces are all triangles. The height of cones and pyramids is the perpendicular distance from the apex to the base. The slant height is the distance along the lateral surface perpendicular to the perimeter of the base.

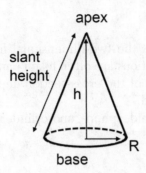

The volume of pyramids and cones are found from the same formula:

$V = \frac{1}{3}Bh$ where B is the area of the base.

SPHERES

A *sphere* is the set of points a fixed distance from a center point, and its volume is given by the formula $V = \frac{4}{3}\pi R^3$.

SURFACE AREA AND LATERAL AREA

The *surface area* is the area of all faces of a solid. Surface area can be found by calculating the area of all the faces of a solid individually and then summing them. You should be able to calculate the surface area of cubes, prisms, and pyramids since the faces of these solids are all polygons. Surface area of cones, cylinders, and spheres involve curved surfaces and are outside the scope of this course.

The lateral area of a solid excludes the bases. For a prism, do not include the two parallel bases. For a pyramid, exclude the one base.

Lateral face—any face of a solid other than its bases
Lateral area—the area of all the lateral faces of a solid
Surface area—the total area of a solid, lateral area + area of bases

CROSS-SECTIONS

A cross-section of a solid is the two-dimensional figure created when a plane intercepts a solid. Cross-sections are often taken parallel or perpendicular to the base of a figure. The shape of the cross-section depends on the angle at which the plane intersects the solid.

Cross-sections of a pyramid, sphere, and cylinder are shown in the accompanying figure.

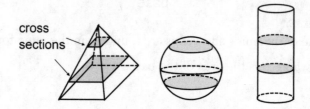

A cross-section of a pyramid taken perpendicular to the base will be shaped triangular, as shown in the figure, or trapezoidal. The cross-section of a cylinder perpendicular to the base is a rectangle.

SOLIDS OF REVOLUTION

Some solids can be generated by rotating a planar figure around a line. These are called *solids of revolution*. Some examples of solids of revolution and the figures that generate them are given in the following table.

Figure Rotated	Solid
Right triangle, rotate 360° about leg $\overline{AB}$	Cone 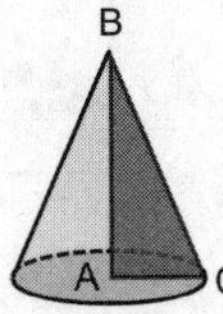
Rectangle $ABCD$, rotate 360° around side $\overline{CD}$	Cylinder 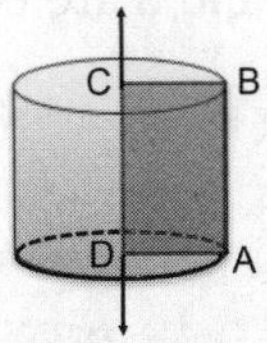
Circle P, rotate 180° about diameter $\overline{AB}$	Sphere 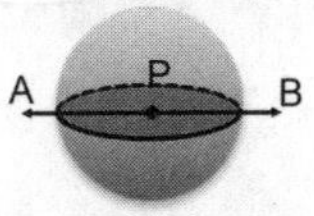

CAVALIERI'S PRINCIPLE

If two solids are contained between two parallel planes, and every parallel plane between these two planes intercepts regions of equal area, then the solids have equal volume. Also, any two parallel planes intercept two solids of equal volume.

Both the triangular prism and the rectangular prism shown in the accompanying figure are bounded by planes R and T. "Bounded" means that the solids have at least one point contained in each of planes R and T, and that the solids do not pass through to the other sides of the planes. As plane S sweeps upward from R toward T, corresponding cross-sections are formed in the triangular prism and the rectangular prism. If every pair of corresponding cross-sections are equal in area (though not necessarily congruent), then the two solids have the same volume.

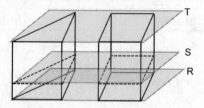

The second part of Cavalieri's principle is illustrated by parallel planes *R* and *S*. If the conditions of Cavalieri's principle are true, then the two volumes between planes *R* and *S* are equal.

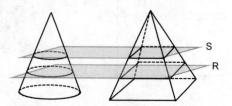

Example

Two stacks of quarters are shown in the accompanying figure. If each stack contains 7 identical quarters, what must be true about the volume of the two stacks? Explain your answer in terms of Cavalieri's principle.

Solution:

Each of the quarters has a uniform and congruent cross-section, and any plane parallel to the bases of the two stacks must intercept cross-sections of equal area. Cavalieri's principle states that the volumes of the stacks must be equal.

JUSTIFICATION OF THE AREA AND VOLUME FORMULAS

Circumference of a Circle

The formula for the area of a circle can be derived using a limit argument by considering a polygon inscribed in a circle and letting the number of sides increase toward infinity.

Consider a polygon inscribed in circle O, as shown in the figure. The central angle, $\angle AOB$, has a measure equal to $\dfrac{360°}{n}$, where n is the number of sides. Altitude $\overline{OP}$ will form $\triangle AOP$, where $m\angle AOP$ equals half the central angle, or $\left(\dfrac{360°}{2n}\right)$, and AO has a length equal to the radius of the circle, R. Length AP can be found by applying the sin ratio.

$$\sin(\theta) = \frac{\text{opposite}}{\text{hypotenuse}}$$

$$\sin\left(\frac{360°}{2n}\right) = \frac{AP}{R}$$

$$AP = R\sin\left(\frac{360°}{2n}\right)$$

The perimeter of the polygon is equal to $2n \cdot AP$, since there are n sides and AP is equal to half a side length.

$$\text{Perimeter} = 2n \cdot AP$$

$$= 2nR\sin\left(\frac{360°}{2n}\right)$$

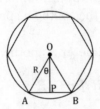

In the limit as the number of sides of the polygon approaches infinity, n approaches infinity, and the perimeter approaches the circumference of the circle. The accompanying table shows the value $n\sin\left(\dfrac{360°}{2n}\right)$ for increasing values of n. As n approaches infinity, $n\sin\left(\dfrac{360°}{2n}\right)$ approaches π.

number of sides n	$n \sin\left(\dfrac{360°}{2n}\right)$
6	3
20	3.1286
100	3.1410
1000	3.1415
Infinite	π

The previous equation for perimeter can be rearranged to

$$\text{Perimeter} = 2n \sin\left(\frac{360°}{2n}\right)R$$

- Substituting circumference for perimeter and π for $n \sin\left(\dfrac{360°}{2n}\right)$ gives the familiar result, circumference $= 2\pi R$

Area of a Circle

The area of a circle can also be derived from a limit argument and inscribed polygons. The formula for the area of a polygon is $A = \frac{1}{2}$ perimeter $\times$ apothem. As the number of sides increases toward infinity, the perimeter approaches the circumference of the circle, and the apothem approaches the radius of the circle.

$$A = \frac{1}{2}\text{perimeter} \times \text{apothem} \rightarrow \frac{1}{2}(2\pi R)R$$

$$A_{\text{circle}} = \frac{1}{2}(2\pi R)R = \pi R^2$$

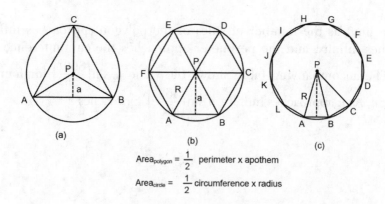

(a)   (b)   (c)

$$\text{Area}_{\text{polygon}} = \frac{1}{2} \text{ perimeter x apothem}$$

$$\text{Area}_{\text{circle}} = \frac{1}{2} \text{ circumference x radius}$$

Volume of a Cylinder

The formula for volume of a cylinder can be derived from the formula for the volume of a prism. As the number of sides of the base of the prism increases toward infinity, the base takes the shape of a circle with area πR^2 resulting in the cylinder formula of $\pi R^2 h$.

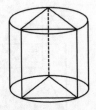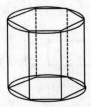

Volume of a Pyramid

The volume of a square-based pyramid can be derived by *dissection* of a cube with side length s, shown in the accompanying figure (a similar method can be used for pyramids of other bases). Let point P be the center of the cube. The four bottom vertices and point P form a square-based pyramid. Five more congruent pyramids can be formed using point P and four vertices at the top, front, back, left, and right. None of the pyramids overlap, so the volume of the cube must be equal to the volume of the 6 pyramids:

$$6 \, \text{Volume}_{\text{pyramid}} = \text{Volume}_{\text{cube}} = s^3$$

$$\text{Volume}_{\text{pyramid}} = \frac{1}{6}s^3$$

The height of each pyramid is equal to $\frac{1}{2}s$, or $s = 2h$. Substitute to get

$$\begin{aligned}
\text{Volume}_{\text{pyramid}} &= \frac{1}{6}s^3 \\
&= \frac{1}{6}s^2 \cdot s \\
&= \frac{1}{6}\left(s^2\right)2h \\
&= \frac{1}{3}\left(s^2\right)h
\end{aligned}$$

The area of the base of each pyramid is the same as the area of the base of the cube, or s^2, so s^2 can be replaced with B to give:

$$\text{Volume}_{\text{pyramid}} = \frac{1}{3}Bh$$

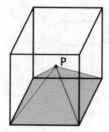 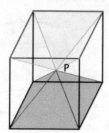

Volume of a Cone

The same approach that was used for the volume of a cylinder can be applied to derive the formula for the volume of a cone. Starting with a pyramid, whose volume is $\frac{1}{3}Bh$, let the number of sides in the polygonal base approach infinity. The shape of the base will approach a circle, but the height will be unchanged, so the formula remains the same:

$$V_{\text{cone}} = \frac{1}{3}Bh$$

MODELING WITH SOLIDS

Many physical objects can be modeled using the basic solids discussed in this chapter. Some examples are:

Cylinders—trees, cans, barrels, people
Spheres—balls, balloons, planets
Prisms—bricks, boxes, aquariums, swimming pools, rooms, books

Volumes or areas can also be used to calculate some other quantity. Some common relationships to look for are

- Mass = density × volume
- Total cost = cost per unit volume × volume

     or

Total cost = cost per unit area × area
- Energy contained in a material = heat content × volume
- Total population = population density × area

For example, if a can has a volume of 100 cm^3 and is filled with a liquid whose density is 1.2 grams/cm^3, then the mass of the liquid inside the can is

$$\text{Mass} = \text{density} \cdot \text{volume}$$
$$\text{Mass} = 1.2 \text{ grams/cm}^3 \cdot 100 \text{ cm}^3$$
$$= 120 \text{ grams}$$

In geometry design problems, we are asked to find the dimensions of some figure or solid that lets the object meet some given condition. Conditions that may be specified include

- maximum or minimum size
- a specified size
- maximum or minimum cost

Approach these problems in the same way as any other modeling problem. The difference comes in using the model to find a maximum or minimum. For example, you may have an equation that gives volume in terms of a length. A maximum or minimum can be found by

- *Trial and error/table*—Using a best first guess for the variable, evaluate the quantity. Change the value of the variable and evaluate the quantity again—it will increase or decrease. Continue until you see the quantity change from decreasing to increasing (a minimum) or increasing to decreasing (a maximum). The table feature of the graphing calculator is excellent for calculating many trials quickly. The table increment can be set to the desired precision.
- *Graphing*—The equation that described the quantity can be graphed, and the maximum/minimum feature of the calculator can then be applied to the graph.
- *Axis of symmetry*—If the equation that describes the quantity is *quadratic* in the form $y = ax^2 + bx + c$, then the maximum or minimum will always occur at the axis of symmetry $x = -b/2a$.

Practice Exercises

1 A basketball has a diameter of 9.6 inches. What is its volume?

   (1) 72.3 in^3 (3) 463 in^3

   (2) 347 in^3 (4) 3706 in^3

2 A right circular cylinder has a volume of 200 cm^3. If the height of the cylinder is 6 cm, what is the radius of the cylinder? Round your answer to the nearest hundredth.

   (1) 2.9 cm   (3) 3.3 cm

   (2) 3.0 cm   (4) 3.8 cm

3 Jenna is preparing to plant a new lawn and has a pile of topsoil delivered to her house. A dump truck delivers the topsoil and dumps it in a pile that is cone shaped. The radius of the pile is 9 ft, and the height is 3 ft. If the topsoil has a density of 95 lb/ft^3, what is the weight of the topsoil?

   (1) 18,764 lb   (3) 28,546 lb

   (2) 24,175 lb   (4) 32,676 lb

4 A right circular cylinder has a radius of 4 and a height of 10. What is the area of a cross-section taken parallel to, and halfway between, the bases? Write your answer in terms of π.

   (1) 4π (3) 12π

   (2) 8π (4) 16π

5 A cylinder and a cone have the same volume. If each solid has the same height, what is the ratio of the radius of the cylinder to the radius of the cone?

(1) 2:1

(2) 3:1

(3) $\sqrt{2}:2$

(4) $\sqrt{3}:3$

6 $\triangle JKL$ is a right triangle with a right angle at K, $JK = 6$, and $KL = 10$. Which of the following solids is generated with $\triangle JKL$ rotated 360° about KL?

(1) a cylinder with a height of 10 and a diameter of 12

(2) a cylinder with a height of 6 and a diameter of 20

(3) a right circular cone with a height of 10 and a diameter of 12

(4) a right circular cone with a height of 6 and a diameter of 20

7 Solids A and B, shown in the figure below, each have uniform cross-sections perpendicular to the shaded bases.

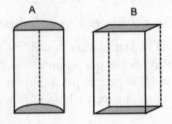

If the heights of the two solids are equal, which of the following statements represents an application of Cavalieri's principle?

(1) The surface areas of solid A and solid B may be different even though their volumes are equal.

(2) Solid A can be generated by rotating a rectangle, whereas solid B is not a solid of revolution.

(3) If the area of base A equals the area of base B, then the volumes of the solids are equal.

(4) If the volume and height of solid A and B are equal, then their surface areas must be equal.

8 The base of a cone is described by the curve $x^2 + y^2 = 16$. If the altitude of the cone intersects the center of its base, and the volume of the cone is 72π, what is the height of the cone?

9 A cylindrical piece of metal has a radius of 15 in and a height of 2 in. It goes through a hot-press machine that reduces the height of the cylinder to $\frac{1}{4}$ in. What is the new radius of the cylinder, assuming no material is lost?

10 Jack is making a scaled drawing of the floor plan of his home. The scale factor is 1 in:4 ft. The drawing of his living room is a rectangle measuring 5 inches by 3 inches. He is planning to purchase new carpet for the living room that costs $4 per square foot. How much will the carpet cost?

(1) $720

(2) $840

(3) $960

(4) $1,040

11 The city of Deersfield sits along the bank of the Fox Run River. A map of the city is shown below. The downtown region, indicated by region A, is $\frac{1}{2}$ mile wide and 1 mile long. The population density of Deersfield, except for the downtown region, is 800 people per square mile. The population density of the downtown region is 4,000 people per square mile. What is the population of Deersfield? Round to the nearest whole number.

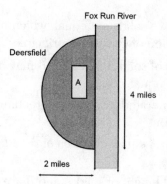

12 An engineer for a construction company needs to calculate the total volume of a home that is under construction so the proper sized heating and air conditioning equipment can be installed. The house is modeled as a rectangular prism with a triangular prism for the roof as shown in the accompanying figure. The roof is symmetric with $EF = ED$. The dimensions are $AB = 30$ ft, $BC = 50$ ft, $BD = 25$ ft, and the measure of $\angle EDF = 40°$. What is the total volume of the house? Round to the nearest cubic foot.

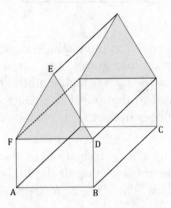

13 The Red Ribbon Orchard sorts their apple harvest by size. The three size categories are shown in the accompanying table.

	Grade B	**Grade A**	**Grade AA**
Average diameter	3 inches	4 inches	5 inches

After one day's harvest, there were 3,000 pounds of grade B apples, 4,200 pounds of grade A apples, and 3,200 pounds of grade AA apples. If the average density of an apple is 0.016 lb/in^3, what is the total number of apples that were harvested?

14 A construction company is working on plans for a project that call for digging a straight 0.25 mile tunnel through a mountain. The cross-section of the tunnel is shown in the accompanying figure.

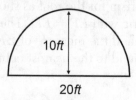

As the workers dig through the mountain, the rubble is brought to a gravel company to be processed into gravel. The cargo beds of the dump trucks are 18 ft long by 10 ft wide by 6 ft high, and the trucking company charges $450 per load delivered to the gravel company. Approximately how much money should the engineer budget for trucking costs for the project?

15 An engineer is designing a hinge to be used on the landing gear of a new airplane. She sketches the cross-section of the hinge on a computer, which is shown in the figure below. The circle represents a hole for a hinge pin.

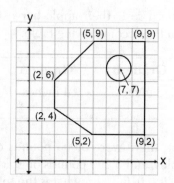

The engineer next uses a function on the computer that creates a solid by translating the cross-section in a direction perpendicular to the sketch. The engineer enters a translation distance of 10 units. In the final step, the engineer enters a scale factor of 1 unit = 2 centimeters. The resulting computer image of the solid is then sent to a factory to be manufactured out of a high-strength metal. If the density of the metal is 8.44 grams/cm^3 and the metal costs $32 per kilogram, what is the expected cost for the metal to make one hinge? Round to the nearest dollar.

Solutions

1 The basketball can be modeled as a sphere whose volume is $\frac{4}{3}\pi R^3$. The radius equals half the diameter, or 4.8 in.

$$V = \frac{4}{3}\pi R^3$$

$$= \frac{4}{3}\pi (4.8 \text{ in})^3$$

$$= 463 \text{ in}^3$$

The correct choice is **(3)**.

2 Start with the formula for volume, substitute the known values, and then solve for the radius.

$$V = B \cdot h$$

$$V = \pi R^2 h$$

$$200 = \pi (R^2)(6)$$

$$R^2 = 10.61032$$

$$R = \sqrt{10.61032}$$

$$= 3.3 \text{ cm}$$

The correct choice is **(3)**.

3 To find the weight of the topsoil, we need to first find the volume of the pile. The volume of a cone is $\frac{1}{3}Bh$, where B is the area of the base. This requires finding the area of the circular base.

$$B = \pi R^2$$

$$= \pi (9^2)$$

$$= 254.4690 \text{ ft}^2$$

We are now ready to find the volume.

$$V = \frac{1}{3}B \cdot h$$

$$= \frac{1}{3}(254.4690)(3)$$

$$= 254.469 \text{ ft}^3$$

Now calculate the weight using the volume and density.

$$\text{weight} = V \cdot \text{density}$$

$$= 254.469 \text{ ft}^3 \cdot 95 \text{ lb/ft}^3$$

$$= 24,174.55 \text{ lb}$$

$$= 24,175 \text{ lb}$$

The correct choice is (**2**).

4 The cross-section will be congruent to the base, so it is a circle with radius of 4. The area is πR^2, or 16π.

The correct choice is (**4**).

5 The strategy is to set the two volume formulas equal to each other and rearrange, solving for the ratio of the radii.

$$V_{\text{cone}} = \frac{1}{3}\pi R_{\text{cone}}^2$$

$$V_{\text{cylinder}} = \pi R_{\text{cylinder}}^2 h$$

$$V_{\text{cylinder}} = V_{\text{cone}}$$

$$\pi R_{\text{cylinder}}^2 h = \frac{1}{3}\pi R_{\text{cone}}^2 h$$

Now solve for the ratio $\dfrac{R_{\text{cylinder}}}{R_{\text{cone}}}$

$$R_{\text{cylinder}}^2 = \frac{1}{3} R_{\text{cone}}^2 \qquad \text{divide by } \pi h$$

$$\frac{R_{\text{cylinder}}^2}{R_{\text{cone}}^2} = \frac{1}{3} \qquad \text{divide by } R_{\text{cone}}^2$$

$$\frac{R_{\text{cylinder}}}{R_{\text{cone}}} = \frac{1}{\sqrt{3}} \qquad \text{take square root of each side}$$

The ratio does not match any of the choices, so try rationalizing the denominator. To do so, multiply the numerator and denominator $\sqrt{3}$.

$$\frac{R_{\text{cylinder}}}{R_{\text{cone}}} = \frac{1}{\sqrt{3}} \cdot \frac{\sqrt{3}}{\sqrt{3}}$$

$$\frac{R_{\text{cylinder}}}{R_{\text{cone}}} = \frac{\sqrt{3}}{3}$$

The correct choice is **(4)**.

6 A right triangle rotated 360° about one of its legs will generate a cone.

The leg aligned with the axis of rotation becomes the height, so the height equals 10. The leg perpendicular to the axis of rotation becomes the radius, so the radius equals 6 and the diameter equals 12.

The correct choice is **(3)**.

7 Cavalieri's principle states that two parallel planes will intercept the same volume in two solids if the cross-sectional areas are uniform and equal. Choice 3 represents this principle.

The correct choice is (**3**).

8 $x^2 + y^2 = 16$ describes a circle centered at the origin with a radius of 4. Its area is

$$A = \pi R^2 = 16\pi$$

Apply the volume formula of the cone to find the height.

$$V = \frac{1}{3}Bh$$

$$72\pi = \frac{1}{3}16\pi h$$

$$h = 72 \cdot 3 \cdot \frac{1}{16}$$

$$= 13.5$$

9 Find the volume of the cylinder before pressing and set it equal to the expression for the volume after pressing. Use this equation to solve for the radius after pressing.

$$V = \pi R^2 h$$

$$V_{before} = \pi \cdot 15^2 \cdot 2$$

$$= 450\pi \text{ in}^3$$

$$V_{after} = \pi \cdot R^2 \cdot \frac{1}{4} \quad \text{we don't know the new radius}$$

$$V_{before} = V_{after}$$

$$\pi \cdot R^2 \cdot \frac{1}{4} = 450\pi \qquad \text{volumes must be equal}$$

$$\frac{R^2}{4} = 450$$

$$R^2 = 1{,}800$$

$$R = \sqrt{1{,}800}$$

$$= 42.426 \text{ in}$$

10 The actual length and width of the living room are found by applying the scale factor to the drawing dimensions.

$$5 \text{ inches} \cdot \frac{4 \text{ ft}}{1 \text{ in}} = 20 \text{ ft}$$

$$3 \text{ inches} \cdot \frac{4 \text{ ft}}{1 \text{ in}} = 12 \text{ ft}$$

The area of the rectangular living room is

$$A = \text{length} \cdot \text{width}$$
$$= 20 \text{ ft} \cdot 12 \text{ ft}$$
$$= 240 \text{ ft}^2$$

The cost of the carpet is

$$\text{cost} = \text{area} \cdot \text{cost per ft}^2$$
$$\text{cost} = 240 \text{ ft}^2 \cdot \frac{\$4}{\text{ft}^2}$$
$$= \$960$$

The correct choice is (**3**).

11 We can model the town as a semicircle and the downtown region as a rectangle. To calculate the population, we first find the area of each region. Let the area of the downtown region be represented by A_A and the area of the remainder of the town be represented by A_B.

$$A_A = \text{length} \cdot \text{width area of circle}$$
$$= \frac{1}{2} \cdot 1$$
$$= 0.5 \text{ mi}^2$$

The area of the remainder of the town is a semicircle minus the area of the downtown region.

$$A_B = \frac{1}{2}\pi R^2 - 0.5 \qquad \text{area of semicircle circle} - \text{area of rectangle}$$
$$= \frac{1}{2}\pi (2)^2 - 0.5$$
$$= 5.78318 \text{ mi}^2$$

The population of each region is the product of the area and the population density

$$\text{Population} = A_\text{A} \cdot 800 \text{ people/mi}^2 + A_\text{B} \cdot 4{,}000 \text{ people/mi}^2$$
$$= 5.78318 \text{ mi}^2 \cdot 800 \text{ people/mi}^2 + 0.5 \text{ mi}^2 \cdot 4{,}000 \text{ people/mi}^2$$
$$= 6{,}627 \text{ people.}$$

The population of Deerfield is 6,627 people.

12 Calculate the volume of each section.

Rectangular prism:

$$\text{length} = BC \quad \text{width} = AB \quad \text{height} = BD$$
$$\text{length} = 50 \text{ ft} \quad \text{width} = 30 \text{ ft} \quad \text{height} = 25 \text{ ft}$$

$$\text{Volume}_\text{rectangular prism} = \text{length} \cdot \text{width} \cdot \text{height}$$
$$= (50 \text{ ft})(30 \text{ ft})(25 \text{ ft})$$
$$= 37{,}500 \text{ ft}^3$$

Triangular prism:

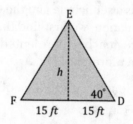

The first step is to find the area of the triangular base, $\triangle FED$. FD has a length equal to AB, or 30 ft. The height of the isosceles triangle divides the base into two 15 ft lengths. The tangent ratio can be used to find the height of $\triangle FED$.

$$\tan = \frac{\text{opposite}}{\text{adjacent}}$$

$$\tan(40°) = \frac{h}{15}$$

$$h = 15\tan(40°)$$

$$= 12.586 \text{ ft}$$

Now multiply the formula for the area of a triangle.

$$A_{\triangle FED} = 2 \cdot \frac{1}{2}\text{base} \cdot \text{height}$$

$$= 15 \text{ ft} \cdot 12.586 \text{ ft}$$

$$= 188.79 \text{ ft}^2$$

The volume of a triangular prism is equal to Bh, where B is the area of the triangular base and h is the length of the prism, which is equal to BC.

$$V_{\text{triangular prism}} = Bh$$

$$= A_{\triangle FED} \cdot BC$$

$$= 188.79 \text{ ft}^2 \cdot 50 \text{ ft}$$

$$= 9{,}439.5 \text{ ft}^3$$

Sum the two volumes to get the total volume of the house.

$$\text{Volume} = 37{,}500 \text{ ft}^3 + 9{,}439 \text{ ft}^3$$

$$= 46{,}939.5 \text{ ft}^3$$

$$= 46{,}940 \text{ ft}^3$$

13 The number of apples of each type equals the total volume of that type divided by the volume of one apple. Model the apples as spheres to calculate the volume of one apple. Find the total volume by dividing the weight by the density. The following table summarizes the calculations.

	Grade B	**Grade A**	**Grade AA**
volume of 1 apple = $V = \frac{4}{3}\pi R^3$	$= \frac{4}{3}\pi \cdot (1.5)^3$ $= 14.1372 \text{ in}^3$	$= \frac{4}{3}\pi \cdot (2)^3$ $= 33.5103 \text{ in}^3$	$= \frac{4}{3}\pi \cdot (2.5)^3$ $= 65.4498 \text{ in}^3$
Total volume of each type = $\frac{\text{weight}}{\text{density}}$	$= \frac{3,000 \text{ lb}}{0.016 \text{ lb/in}^3}$ $= 187,500 \text{ in}^3$	$= \frac{4,200 \text{ lb}}{0.016 \text{ lb/in}^3}$ $= 262,500 \text{ in}^3$	$= \frac{3,000 \text{ lb}}{0.016 \text{ lb/in}^3}$ $= 200,000 \text{ in}^3$
number of each type = $\frac{V_{\text{total}}}{V_{\text{one apple}}}$	$= \frac{187,500}{14.1372}$ $= 13,263$	$= \frac{262,500}{33.5013}$ $= 7,833$	$= \frac{200,000}{65.4498}$ $= 3,056$

The total number of apples is $13,263 + 7,833 + 3,056 = 24,152$ apples.

14 First, find the volume of the tunnel, and divide by the volume of the dump truck bed to find the number of dump-truck loads needed. The cross-section of the tunnel is a semicircle, which makes the tunnel in the shape of a half cylinder. The cylinder radius is 10 ft and the length is 0.25 miles.

First, convert 0.25 miles to feet using the conversion factor from the formula sheet.

$$0.25 \text{ mi} \cdot \frac{5,280 \text{ ft}}{\text{mi}} = 1,320 \text{ ft}$$

$$V_{\text{tunnel}} = \frac{1}{2}\pi R^2 h$$

$$= \frac{1}{2}\pi \cdot 10^2 \cdot 1,320$$

$$= 207,345.11 \text{ ft}^3$$

Model the dump-truck bed as a rectangular prism.

$$V_{truck} = \text{length} \cdot \text{width} \cdot \text{height}$$
$$= 18 \cdot 10 \cdot 6 = 1{,}080 \text{ ft}^3$$

The number of dump-truck loads is the ratio of the two volumes.

Number of loads $= V_{tunnel}/V_{truck}$

$$= \frac{207{,}345.11}{1{,}080}$$
$$= 191.98 \text{ dump-truck loads}$$

192 dump-truck loads are needed.

The estimated cost $= \dfrac{\$450}{\text{load}} \cdot 192 \text{ loads} = \$86{,}400$

15 The solid produced will be a prism with the cross-section shown in the figure. The first step is to calculate the area of the cross-section. Divide the cross-section into the 5 regions shown below. The 5th region represents the hole and must be subtracted from the other areas.

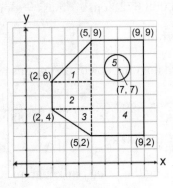

$A_1 = \frac{1}{2}b \cdot h$	$A_2 = l \cdot w$	$A_3 = \frac{1}{2}b \cdot h$	$A_4 = l \cdot w$	$A_5 = \pi R^2$
$= \frac{1}{2} \cdot 3 \cdot 3$	$= 3 \cdot 2$	$= \frac{1}{2} \cdot 3 \cdot 2$	$= 4 \cdot 7$	$= \pi(1)^2$
$= 4.5$	$= 6$	$= 3$	$= 28$	$= 3.14159$

$$A_{\text{drawing}} = A_1 + A_2 + A_3 + A_4 - A_5$$
$$= 4.5 + 6 + 3 + 28 - 3.14159 = 38.35841 \text{ units}^2$$

To find the cross-sectional area of the actual part in centimeters, use the scale factor and the fact that area is proportional to the scale factor *squared*.

$$A_{\text{cross-section}} = 38.35841 \text{ units}^2 \cdot \left(\frac{2 \text{ cm}}{1 \text{ unit}}\right)^2$$
$$= 153.43364 \text{ cm}^2$$

The solid is formed by translating the cross-section 10 units, or 20 cm. Use these dimensions to find the volume of the prism.

$$V = B \times h \qquad \text{where } B \text{ is the cross-sectional area}$$
$$= 153.43364 \text{ cm}^2 \cdot 20 \text{ cm}$$
$$= 3{,}068.6728 \text{ cm}^3$$

Now calculate the mass of the metal

$$= 8.44 \frac{\text{grams}}{\text{cm}^3} \cdot 3{,}068.6728 \text{ cm}^3 \qquad \text{mass} = \text{density} \times \text{volume}$$
$$= 25{,}899.5984 \text{ grams}$$

The price is given per kilogram, so convert grams to kilograms.

$$\text{Mass} = 25{,}899.5984 \text{ grams} \cdot \frac{1 \text{ kg}}{1{,}000 \text{ grams}}$$
$$= 25.899 \text{ kg}$$

Finally, find the cost using the price per kilogram.

$$\text{Cost} = 25.899 \text{ kg} \cdot \$32/\text{kg} \quad \text{total cost} = \text{mass (price per kg)}$$
$$= \$829$$

Glossary of Geometry Terms

Acute angle An angle whose measure is greater than 0° and less than 90°.

Acute triangle A triangle whose angles are all acute angles.

Adjacent angles Two angles that share a common vertex and one side, but do not share interior points.

Altitude (of a triangle) A segment from a vertex of a triangle perpendicular to the opposite side.

Angle A figure formed by two rays with a common endpoint. Symbol is $\angle$.

Angle bisector A line, segment, or ray that divides an angle into two congruent angles.

Angle measure The amount of opening of an angle, measured in degrees or radians.

Angle of depression The angle formed by the horizontal and the line of sight when looking downward to an object.

Angle of elevation The angle formed by the horizontal and the line of sight when looking upward to an object.

Angle of rotation The angle measure by which a figure or point spins around a center point.

Arc A portion of a circle, with two endpoints on the circle. A **major arc** measures more than 180°, a **minor arc** measures less than 180°, and a **semicircle** measures 180°. Symbol ⌒.

Arc length The distance between the endpoints of an arc. Arc length equals $R\theta$ where θ is the arc measure in radians and R is the radius.

Arc measure The angle measure of an arc equal to the measure of the central angle that intercepts the arc.

Base (of a circular cone) The circular face of a cone; it is opposite the apex.

Base (of an isosceles triangle) The noncongruent side of an isosceles triangle.

Base (of a pyramid) The polygonal face of a pyramid that is opposite the apex.

Bases (of a prism) A pair of faces of a prism that are parallel, congruent polygons.

Bases (of a circular cylinder) The pair of congruent parallel circular faces of a cylinder.

Base angles The angles formed at each end of the base of an isosceles triangle.

Bisector, or segment bisector A line, segment, or ray that passes through the midpoint of a segment.

Biconditional A compound statement in the form "If and only if *hypothesis* then *conclusion.*" The hypothesis and conclusion are statements that can be true or false. The biconditional is true when the hypothesis and conclusion have the same truth value.

Cavalieri's principle If two solids are contained between two parallel planes, and every parallel plane between these two planes intercepts regions of equal area, then the solids have equal volume. Also, any two parallel planes intercept two solids of equal volume.

Center of dilation The fixed reference point used to determine the expansion or contraction of lengths in a dilation. The center of dilation is the only invariant point in a dilation.

Center of a regular polygon The center of the inscribed or circumscribed circle of a regular polygon.

Center–radius equation of a circle A circle can be represented on the coordinate plane by $(x - h)^2 + (y - k)^2 = r^2$, where the point (h, k) is the center of the circle and r is the radius.

Central angle An angle in a circle formed by two distinct radii.

Centroid of a triangle The point of concurrency of the three medians of a triangle.

Chord A segment whose endpoints lie on a circle.

Circle The set of points that are a fixed distance from a fixed center point. Symbol $\odot$.

Circumcenter The point that is the center of the circle circumscribed about a polygon, equidistant from the vertices of a polygon, and the point of concurrency of the perpendicular bisectors of a triangle.

Circumference The distance around a circle. Circumference $= 2\pi \cdot$ radius.

Coincide (coincident) Figures that lay entirely one on the other.

Collinear Points that lie on the same line.

Compass A tool for drawing accurate circles and arcs.

Complementary angles Angles whose measures sum to 90°.

Completing the square A method used to rewrite a quadratic expression of the form $ax^2 + bx + c$ as a squared binomial of the form $(dx + e)^2$.

Composition of transformations A sequence of transformations in a specified order.

Concave polygon A polygon with at least one diagonal outside the polygon.

Concentric circles Circles with the same center.

Concurrent When three or more lines all intersect at a single point.

Cone (circular) A solid figure with a single circular base, and a curved lateral face that tapers to a point called the apex.

Congruent Figures with the same size and shape. Symbol $\cong$.

Construction A figure drawn with only a compass and straightedge.

Convex polygon A polygon whose diagonals all lie within the polygon.

Coplanar Figures that lie in the same plane.

Corresponding parts A pair of parts (usually points, sides, or angles) of two figures that are paired together through a specified relationship, such as a congruence or similarity statement or a transformation function.

Cosine of an angle In a right triangle, the ratio of the length of the side adjacent to an acute angle to the length of the hypotenuse.

CPCTC Corresponding parts of congruent triangles are congruent.

Cross-section The intersection of a plane with a solid.

Cube A prism whose faces are all squares.

Cubic unit The amount of space occupied by a cube 1 unit of length in each dimension. (1 ft^3 is the amount of space occupied by a 1 ft $\times$ 1 ft $\times$ 1 ft cube.)

Cylinder The solid figure formed when a rectangle is rotated 360° about one of its sides.

Cylinder (circular) A solid with two parallel, congruent circular bases and a curved lateral face.

Degree A unit of angle measure equal to $\frac{1}{360}$th of a complete rotation.

Diagonal A segment in a polygon whose endpoints are nonconsecutive vertices of the polygon.

Diameter A chord that passes through the center of a circle.

Dilation A similarity transformation about a center point O, which maps pre-image P to image P' such that O, P, and P' are collinear and $OP' = k \cdot OP$. Notation is D_k.

Direct transformation A transformation that preserves orientation.

Distance from a point to a line The length of the segment from the point perpendicular to the line.

Edge of a solid The intersection of two polygonal faces of a solid.

Endpoint The two points that bound a segment.

Equiangular A figure whose angles all have the same measure.

Equidistant The same distance from two or more points.

Equilateral A figure whose sides all have the same length.

Exterior angle of a polygon The angle formed by a side of a polygon and the extension of an adjacent side.

Face of a solid Any of the surfaces that bound a solid.

Figure Any set of points. The points may form a segment, line, ray, plane, polygon, curve, solid, etc.

Glide reflection The composition of a line reflection and a translation along a vector parallel to the line of reflection.

Great circle The largest circle that can be drawn on a sphere. The circle formed by the intersection of a sphere and a plane passing through the center of the sphere.

Hexagon A 6-sided polygon.

Hypotenuse The longest side of a right triangle; it is always opposite the right angle.

Identity transformation A transformation in which the pre-image and image are coincident.

Image The figure that results from applying a transformation to an initial figure called the pre-image.

Incenter of a triangle The point that is the center of the inscribed circle of a triangle, equidistant from the 3 sides of a triangle, and the point of concurrency of the three angle bisectors of a triangle.

Inscribed angle An angle in a circle formed by two chords with a common endpoint.

Inscribed circle of a triangle A circle that is tangent to all three sides of the triangle.

Intersecting Figures that share at least one common point.

Isometry See *Rigid motion*.

Isosceles trapezoid A trapezoid with congruent legs.

Isosceles triangle A triangle with at least two congruent sides.

Kite A polygon with two distinct pairs of adjacent congruent sides. The opposite sides are not congruent.

Lateral edge The intersection between two lateral faces of a polyhedron.

Lateral face Any face of a polyhedron that is not a base.

Line One of the undefined terms in geometry. An infinitely long set of points that has no width or thickness. Symbol $\leftrightarrow$.

Linear pair of angles Two adjacent supplementary angles.

Line symmetry A line over which one half of a figure can be reflected and mapped onto the other half of the figure.

Line reflection A rigid motion in which every point P is transformed to a point P' such that the line is the perpendicular bisector of $\overline{PP'}$.

Major arc An arc whose degree measure is greater than 180°.

Map (mapping) A pairing of every point in a pre-image with a point in an image. Reflections, rotations, translations, and dilations are one-to-one mappings because every point in the pre-image maps to exactly one point in the image, and every point in the image maps to exactly one point in the pre-image.

Mean proportional (geometric mean) The square root of the product of two numbers, a and b. If $\dfrac{a}{m} = \dfrac{m}{b}$, then m is the geometric mean.

Median of a triangle The segment from a vertex of a triangle to the midpoint of the opposite side.

Midpoint A point that divides a segment into two congruent segments.

Midsegment (median) of a trapezoid A segment joining the midpoints of the two legs of a trapezoid.

Midsegment of a triangle A segment joining the midpoints of two sides of a triangle.

Minor arc An arc whose degree measure is less than 180°.

Noncollinear Points that do not lie on the same line.

Noncoplanar Points or lines that do not lie on the same plane.

Obtuse angle An angle whose measure is greater than 90° and less than 180°.

Octagon A polygon with 8 sides.

Opposite rays Two rays with a common endpoint that together form a straight line.

Opposite transformation A transformation that changes the orientation of a figure.

Orientation The order in which the points of a figure are encountered when moving around a figure. The orientation can be clockwise or counterclockwise.

Orthocenter The point of concurrence of the three altitudes of a triangle.

Parallel lines Coplanar lines that do not intersect. Symbol //.

Parallelogram A quadrilateral with two pairs of opposite parallel sides.

Parallel planes Planes that do not intersect.

Pentagon A polygon with 5 sides.

Perimeter The sum of the lengths of the sides of a polygon.

Perpendicular Intersecting at right angles. Symbol $\perp$.

Perpendicular bisector A line, segment, or ray perpendicular to another segment at its midpoint.

Pi The ratio of the circumference of a circle to its diameter. Pi is an irrational number whose value is approximately 3.14159. Symbol π.

Plane One of the undefined terms in geometry. A set of points with no thickness that extends infinitely in all directions. It is usually visualized as a flat surface.

Point An undefined term in geometry. A location in space with no length, width, or thickness. Symbol $\bullet$.

Point of tangency The point where a tangent intersects a curve.

Point reflection A transformation in which a specified center point is the midpoint of each of the segments connecting any point in the pre-image with its corresponding point in the image. It is equivalent to a rotation of 180°.

Point-slope equation of a line A form of the equation of a line, $y - y_1 = m(x - x_1)$, where m is the slope and (x_1, y_1) are the coordinates of a point on the line.

Polygon A closed planar figure whose sides are segments that intersect only at their endpoints (do not overlap).

Polyhedron A solid figure in which each face is a polygon. Plural *polyhedra*. Prisms and pyramids are examples of polyhedra.

Postulate A statement that is accepted to be true without proof.

Pre-image The original figure that is acted on by a transformation.

Preserves Remains unchanged.

Prism A polyhedron with two congruent, parallel, polygons for bases, and whose lateral faces are parallelograms.

Proportion An equation that states two ratios are equal. For example, $\frac{a}{b} = \frac{c}{d}$. The cross-products of a proportion are equal, $a \cdot d = b \cdot c$.

Pyramid A polyhedron having one polygonal base and triangles for lateral faces.

Pythagorean theorem In a right triangle, the sum of the squares of the two legs equals the square of the hypotenuse, or $a^2 + b^2 = c^2$, where a and b are the lengths of the two legs and c is the length of the hypotenuse.

Quadratic equation An equation in the form $ax^2 + bx + c = 0$, where $a \neq 0$.

Quadratic formula A formula for finding the two solutions to a quadratic equation of the form $ax^2 + bx + c = 0$, $x = \dfrac{-b \pm \sqrt{b^2 - 4ac}}{2a}$.

Quadrilateral A polygon with four sides.

Radian An angle measure in which one full rotation is 2π radians. Also, 1 radian is the measure of an arc such that the arc's length is equal to the radius of that circle.

Radius A segment from the center of a circle to a point on the circle.

Ray A portion of a line starting at an endpoint and including all points on one side of the endpoint. Symbol $\rightarrow$.

Rectangle A parallelogram with right angles.

Reflection See *Line reflection* and *Point reflection*.

Reflexive property of equality Any quantity is equal to itself. Also, for figures, any figure is congruent to itself.

Regular polygon A polygon in which all sides are congruent and all angles are congruent.

Rhombus A parallelogram with 4 congruent sides.

Right angle An angle that measures $90°$.

Right circular cone A cone with a circular base and whose altitude passes through the center of the base.

Right circular cylinder A cylinder with circular bases and whose altitude passes through the center of the bases.

Right pyramid A pyramid whose faces are isosceles triangles.

Right triangle A triangle that contains a right angle.

Rigid motion (isometry) A transformation that preserves distance. The image and pre-image are congruent under a rigid motion. Translations, reflections, and rotations are isometries.

Rotation A rigid motion in which every point, P, in the pre-image spins by a fixed angle around a center point, C, to point P'. The distance to the center point is preserved.

Rotational symmetry A figure has rotational symmetry if it maps onto itself after a rotation of less than $360°$.

Scale drawing A drawing of a figure or object that represents a dilation of the actual figure or object. A drawing of an object in which every length is enlarged or reduced by the same scale factor.

Scale factor The ratio by which a figure is enlarged or reduced by a dilation.

Scalene triangle A triangle in which no sides have the same length.

Secant of a circle A line that intersects a circle in exactly two points.

Sector of a circle A region bounded by a central angle of a circle and the arc it intersects.

Segment A portion of a line bounded by two endpoints.

Semicircle An arc that measures 180°.

Similarity transformation A transformation in which the pre-image and image are similar. A transformation that includes a dilation.

Similar polygons Two polygons with the same shape but not necessarily the same size.

Sine of an angle In a right triangle, the ratio of the length of the side opposite an acute angle to the length of the hypotenuse.

Skew lines Two lines that are not coplanar.

Slant height The distance along a lateral face of a solid from the apex perpendicular to the opposite edge.

Slope A numerical measure of the steepness of a line. In the coordinate plane, the slope of a line equals the change in the y-coordinates divided by the change in the x-coordinates between any two points. The slope of a vertical line is undefined.

Slope-intercept equation of a line A form of the equation of a line, $y = mx + b$, where m is the slope and b is the y-intercept.

Solid figure A 3-dimensional figure fully enclosed by surfaces.

Sphere A solid comprised of the set of points in space that are a fixed distance from a center point.

Square A parallelogram with 4 right angles and 4 congruent sides.

Straightedge A ruler with no length marking, used for constructing straight lines.

Substitution property of equality The property that a quantity can be replaced by an equal quantity in an equation.

Subtraction property of equality If the same or equal quantities are subtracted from the same or equal quantities, then the differences are equal.

Supplementary angles Two angles whose measures sum to 180°.

Surface area The sum of the areas of all the faces or curved surfaces of a solid figure.

Tangent of an angle In a right triangle, the ratio of the length of the side opposite an acute angle to the length of the side adjacent to the angle.

Tangent to a circle A line, coplanar with a circle, that intersects the circle at only one point.

Theorem A general statement that can be proven.

Transformation A one-to-one function that maps a set of points, called the pre-image, to a new set of points, called the image.

Transitive property of congruence The property that states: if figure $A \cong$ figure B and figure $B \cong$ figure C, then figure $A \cong$ figure C.

Transitive property of equality The property that states: if $a = b$ and $b = c$, then $a = c$.

Translation A rigid motion that slides every point in the pre-image in the same direction and by the same distance.

Transversal A line that intersects two or more other lines at different points.

Trapezoid A quadrilateral that has at least one pair of parallel sides.

Undefined terms Terms which cannot be formally defined using previously defined terms. In geometry these traditionally include point, line, and plane.

Vector A quantity that has both magnitude and direction; represented geometrically by a directed line segment. Symbol $\rightarrow$.

Vertex The point of intersection of two consecutive sides of a polygon or the two rays of an angle.

Vertex angle The angles formed by the two congruent sides in an isosceles triangle.

Vertical angles The pairs of opposite angles formed by two intersecting lines.

Volume The amount of space occupied by a solid figure measured in cubic units (in^3, cm^3, etc.). The number of nonoverlapping unit cubes that can fit in the interior of a solid.

Zero product property The property that states: if $a \cdot b = 0$, then either $a = 0$, or $b = 0$, or a and $b = 0$.

Regents Examinations, Answers, and Self-Analysis Charts

Examination
August 2017
Geometry (Common Core)

GEOMETRY REFERENCE SHEET

1 inch = 2.54 centimeters

1 meter = 39.37 inches

1 mile = 5280 feet

1 mile = 1760 yards

1 mile = 1.609 kilometers

1 kilometer = 0.62 mile

1 pound = 16 ounces

1 pound = 0.454 kilogram

1 kilogram = 2.2 pounds

1 ton = 2000 pounds

1 cup = 8 fluid ounces

1 pint = 2 cups

1 quart = 2 pints

1 gallon = 4 quarts

1 gallon = 3.785 liters

1 liter = 0.264 gallon

1 liter = 1000 cubic centimeters

Triangle	$A = \frac{1}{2}bh$
Parallelogram	$A = bh$
Circle	$A = \pi r^2$
Circle	$C = \pi d$ or $C = 2\pi r$

General Prisms	$V = Bh$
Cylinder	$V = \pi r^2 h$
Sphere	$V = \frac{4}{3}\pi r^3$
Cone	$V = \frac{1}{3}\pi r^2 h$
Pyramid	$V = \frac{1}{3}Bh$
Pythagorean Theorem	$a^2 + b^2 = c^2$
Quadratic Formula	$x = \dfrac{-b \pm \sqrt{b^2 - 4ac}}{2a}$
Arithmetic Sequence	$a_n = a_1 + (n-1)d$
Geometric Sequence	$a_n = a_1 r^{n-1}$
Geometric Series	$S_n = \dfrac{a_1 - a_1 r^n}{1 - r}$ where $r \neq 1$
Radians	$1\ \text{radian} = \dfrac{180}{\pi}\ \text{degrees}$
Degrees	$1\ \text{degrees} = \dfrac{\pi}{180}\ \text{radians}$
Exponential Growth/Decay	$A = A_0 e^{k(t - t_0)} + B_0$

180 Minutes–36 Questions

PART I

Answer all 24 questions in this part. Each correct answer will receive 2 credits. No partial credit will be allowed. For each statement or question, write in the space provided the numeral preceding the word or expression that best completes the statement or answers the question. [48 credits]

1 A two-dimensional cross section is taken of a three-dimensional object. If this cross section is a triangle, what can *not* be the three-dimensional object?

(1) cone   (3) pyramid

(2) cylinder   (4) rectangular prism   1 _____

2 The image of $\triangle DEF$ is $\triangle D'E'F'$. Under which transformation will the triangles *not* be congruent?

(1) a reflection through the origin

(2) a reflection over the line $y = x$

(3) a dilation with a scale factor of 1 centered at $(2, 3)$

(4) a dilation with a scale factor of $\frac{3}{2}$ centered at the origin   2 _____

3 The vertices of square $RSTV$ have coordinates $R(-1, 5)$, $S(-3, 1)$, $T(-7, 3)$, and $V(-5, 7)$. What is the perimeter of $RSTV$?

(1) $\sqrt{20}$ (3) $4\sqrt{20}$

(2) $\sqrt{40}$ (4) $4\sqrt{40}$ 3 _____

4 In the diagram below of circle O, chord $\overline{CD}$ is parallel to diameter $\overline{AOB}$ and $m\overparen{CD} = 130°$.

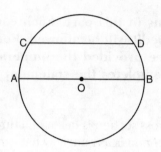

What is $m\overparen{AC}$?

(1) 25°

(2) 50°

(3) 65°

(4) 115°

4 _____

5 In the diagram below, $\overline{AD}$ intersects $\overline{BE}$ at C, and $\overline{AB} // \overline{DE}$.

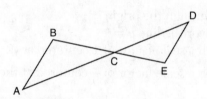

If $CD = 6.6$ cm, $DE = 3.4$ cm, $CE = 4.2$ cm, and $BC = 5.25$ cm, what is the length of $\overline{AC}$, to the *nearest hundredth of a centimeter*?

(1) 2.70

(2) 3.34

(3) 5.28

(4) 8.25

5 _____

6 As shown in the graph below, the quadrilateral is a rectangle.

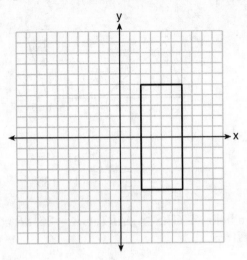

Which transformation would *not* map the rectangle onto itself?

(1) a reflection over the *x*-axis
(2) a reflection over the line $x = 4$
(3) a rotation of 180° about the origin
(4) a rotation of 180° about the point $(4, 0)$ 6 _____

7 In the diagram below, triangle *ACD* has points *B* and *E* on sides $\overline{AC}$ and $\overline{AD}$, respectively, such that $\overline{BE}//\overline{CD}$, $AB = 1$, $BC = 3.5$, and $AD = 18$.

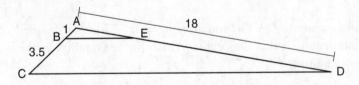

What is the length of $\overline{AE}$, to the *nearest tenth*?

(1) 14.0   (3) 3.3
(2) 5.1   (4) 4.0   7 _____

8 In the diagram below of parallelogram $ROCK$, m$\angle C$ is 70° and m$\angle ROS$ is 65°.

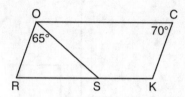

What is m$\angle KSO$?

(1) 45°

(3) 115°

(2) 110°

(4) 135°

8 _____

9 In the diagram below, $\angle GRS \cong \angle ART$, $GR = 36$, $SR = 45$, $AR = 15$, and $RT = 18$.

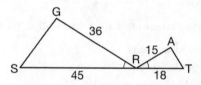

Which triangle similarity statement is correct?

(1) $\triangle GRS \sim \triangle ART$ by AA.

(2) $\triangle GRS \sim \triangle ART$ by SAS.

(3) $\triangle GRS \sim \triangle ART$ by SSS.

(4) $\triangle GRS$ is not similar to $\triangle ART$.

9 _____

10 The line represented by the equation $4y = 3x + 7$ is transformed by a dilation centered at the origin. Which linear equation could represent its image?

(1) $3x - 4y = 9$

(3) $4x - 3y = 9$

(2) $3x + 4y = 9$

(4) $4x + 3y = 9$

10 _____

11 Given $\triangle ABC$ with m$\angle B = 62°$ and side $\overline{AC}$ extended to D, as shown below.

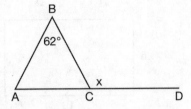

What value of x makes $\overline{AB} \cong \overline{CB}$?

(1) 59°   (3) 118°

(2) 62°   (4) 121°   11____

12 In the diagram shown below, $\overline{PA}$ is tangent to circle T at A, and secant $\overline{PBC}$ is drawn where point B is on circle T.

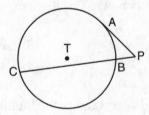

If $PB = 3$ and $BC = 15$, what is the length of $\overline{PA}$?

(1) $3\sqrt{5}$ (3) 3

(2) $3\sqrt{6}$ (4) 9   12____

13 A rectangle whose length and width are 10 and 6, respectively, is shown below. The rectangle is continuously rotated around a straight line to form an object whose volume is 150π.

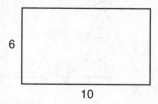

Which line could the rectangle be rotated around?

(1) a long side
(2) a short side
(3) the vertical line of symmetry
(4) the horizontal line of symmetry   13 _____

14 If *ABCD* is a parallelogram, which statement would prove that *ABCD* is a rhombus?

(1) $\angle ABC \cong \angle CDA$ (3) $\overline{AC} \perp \overline{BD}$

(2) $\overline{AC} \cong \overline{BD}$ (4) $\overline{AB} \perp \overline{CD}$ 14 _____

15 To build a handicapped-access ramp, the building code states that for every 1 inch of vertical rise in height, the ramp must extend out 12 inches horizontally, as shown in the diagram below.

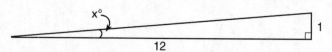

What is the angle of inclination, *x*, of this ramp, to the *nearest hundredth of a degree*?

(1) 4.76   (3) 85.22

(2) 4.78   (4) 85.24   15 _____

16 In the diagram below of △ABC, D, E, and F are the midpoints of
    $\overline{AB}$, $\overline{BC}$, and $\overline{CA}$, respectively.

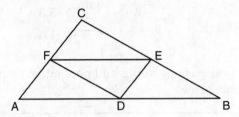

What is the ratio of the area of △CFE to the area of △CAB?

(1) 1 : 1   (3) 1 : 3

(2) 1 : 2   (4) 1 : 4   16 _____

17 The coordinates of the endpoints of $\overline{AB}$ are A(−8, −2) and
    B(16, 6). Point P is on $\overline{AB}$. What are the coordinates of point P,
    such that AP : PB is 3 : 5?

(1) (1, 1)   (3) (9.6, 3.6)

(2) (7, 3)   (4) (6.4, 2.8)   17 _____

18 Kirstie is testing values that would make triangle KLM a right
    triangle when $\overline{LN}$ is an altitude, and KM = 16, as shown below.

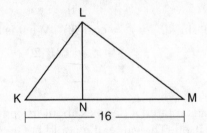

Which lengths would make triangle KLM a right triangle?

(1) LM = 13 and KN = 6   (3) KL = 11 and KN = 7

(2) LM = 12 and NM = 9   (4) LN = 8 and NM = 10 18 _____

19 In right triangle ABC, m$\angle A = 32°$, m$\angle B = 90°$, and $AC = 6.2$ cm. What is the length of $\overline{BC}$, to the *nearest tenth of a centimeter*?

(1) 3.3   (3) 5.3

(2) 3.9   (4) 11.7   19 ____

20 The 2010 U.S. Census populations and population densities are shown in the table below.

State	Population Density $\left(\dfrac{\text{people}}{\text{mi}^2}\right)$	Population in 2010
Florida	350.6	18,801,310
Illinois	231.1	12,830,632
New York	411.2	19,378,102
Pennsylvania	283.9	12,702,379

Based on the table above, which list has the states' areas, in square miles, in order from largest to smallest?

(1) Illinois, Florida, New York, Pennsylvania
(2) New York, Florida, Illinois, Pennsylvania
(3) New York, Florida, Pennsylvania, Illinois
(4) Pennsylvania, New York, Florida, Illinois   20 ____

21 In a right triangle, $\sin(40 - x)° = \cos(3x)°$. What is the value of x?

(1) 10   (3) 20

(2) 15   (4) 25   21 ____

22 A regular decagon is rotated n degrees about its center, carrying the decagon onto itself. The value of n could be

(1) 10°   (3) 225°

(2) 150°   (4) 252°   22 ____

23 In a circle with a diameter of 32, the area of a sector is $\dfrac{512\pi}{3}$. The measure of the angle of the sector, in radians, is

(1) $\dfrac{\pi}{3}$ (3) $\dfrac{16\pi}{3}$

(2) $\dfrac{4\pi}{3}$ (4) $\dfrac{64\pi}{3}$ 23 _____

24 What is an equation of the perpendicular bisector of the line segment shown in the diagram below?

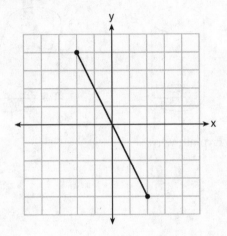

(1) $y + 2x = 0$ (3) $2y + x = 0$

(2) $y - 2x = 0$ (4) $2y - x = 0$ 24 _____

PART II

Answer all 7 questions in this part. Each correct answer will receive 2 credits. Clearly indicate the necessary steps, including appropriate formula substitutions, diagrams, graphs, charts, etc. For all questions in this part, a correct numerical answer with no work shown will receive only 1 credit.   [14 credits]

25 Sue believes that the two cylinders shown in the diagram below have equal volumes.

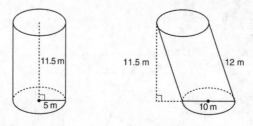

Is Sue correct? Explain why.

26 In the diagram of rhombus *PQRS* below, the diagonals $\overline{PR}$ and $\overline{QS}$ intersect at point *T*, *PR* = 16, and *QS* = 30. Determine and state the perimeter of *PQRS*.

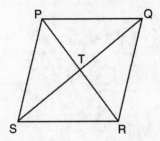

27 Quadrilateral *MATH* and its image *M″A″T″H″* are graphed on the set of axes below.

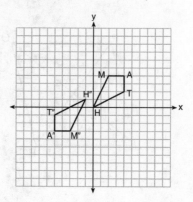

Describe a sequence of transformations that maps quadrilateral *MATH* onto quadrilateral *M″A″T″H″*.

28 Using a compass and straightedge, construct a regular hexagon inscribed in circle *O*.

[Leave all construction marks.]

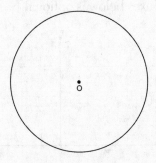

29 The coordinates of the endpoints of $\overline{AB}$ are $A(2, 3)$ and $B(5, -1)$.
Determine the length of $\overline{A'B'}$, the image of $\overline{AB}$, after a dilation of $\frac{1}{2}$ centered at the origin.

[The use of the set of axes below is optional.]

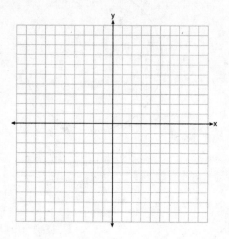

30 In the diagram below of $\triangle ABC$ and $\triangle XYZ$, a sequence of rigid motions maps $\angle A$ onto $\angle X$, $\angle C$ onto $\angle Z$, and $\overline{AC}$ onto $\overline{XZ}$.

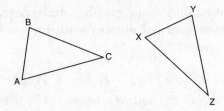

Determine and state whether $\overline{BC} \cong \overline{YZ}$. Explain why.

31 Determine and state the coordinates of the center and the length of the radius of a circle whose equation is $x^2 + y^2 - 6x = 56 - 8y$.

PART III

Answer all 3 questions in this part. Each correct answer will receive 4 credits. Clearly indicate the necessary steps, including appropriate formula substitutions, diagrams, graphs, charts, etc. For all questions in this part, a correct numerical answer with no work shown will receive only 1 credit.   [12 credits]

32 Triangle PQR has vertices $P(-3, -1)$, $Q(-1, 7)$, and $R(3, 3)$, and points A and B are midpoints of $\overline{PQ}$ and $\overline{RQ}$, respectively. Use coordinate geometry to prove that $\overline{AB}$ is parallel to $\overline{PR}$ and is half the length of $\overline{PR}$.

[The use of the set of axes below is optional.]

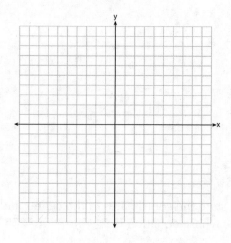

33 In the diagram below of circle O, tangent $\overleftrightarrow{EC}$ is drawn to diameter $\overline{AC}$. Chord $\overline{BC}$ is parallel to secant $\overrightarrow{ADE}$, and chord $\overline{AB}$ is drawn.

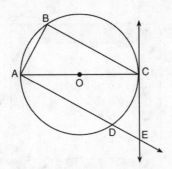

Prove: $\dfrac{BC}{CA} = \dfrac{AB}{EC}$

34 Keira has a square poster that she is framing and placing on her wall. The poster has a diagonal 58 cm long and fits exactly inside the frame. The width of the frame around the picture is 4 cm.

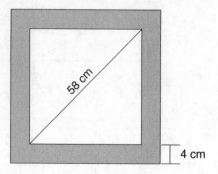

Determine and state the total area of the poster and frame to the *nearest tenth of a square centimeter*.

PART IV

Answer the 2 questions in this part. Each correct answer will receive 6 credits. Clearly indicate the necessary steps, including appropriate formula substitutions, diagrams, graphs, charts, etc. For all questions in this part, a correct numerical answer with no work shown will receive only 1 credit.   **[12 credits]**

35 Isosceles trapezoid $ABCD$ has bases $\overline{DC}$ and $\overline{AB}$ with nonparallel legs $\overline{AD}$ and $\overline{BC}$. Segments AE, BE, CE, and DE are drawn in trapezoid $ABCD$ such that $\angle CDE \cong \angle DCE$, $\overline{AE} \perp \overline{DE}$, and $\overline{BE} \perp \overline{CE}$.

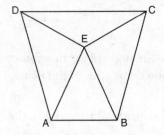

Prove $\triangle ADE \cong \triangle BCE$ and prove $\triangle AEB$ is an isosceles triangle.

36 A rectangular in-ground pool is modeled by the prism below. The inside of
the pool is 16 feet wide and 35 feet long. The pool has a shallow end and a
deep end, with a sloped floor connecting the two ends. Without water, the
shallow end is 9 feet long and 4.5 feet deep, and the deep end of the pool is
12.5 feet long.

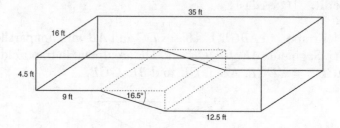

If the sloped floor has an angle of depression of 16.5 degrees, what is the
depth of the pool at the deep end, to the *nearest tenth of a foot*?

Find the volume of the inside of the pool to the *nearest cubic foot*.

A garden hose is used to fill the pool. Water comes out of the hose at a rate
of 10.5 gallons per minute. How much time, to the *nearest hour*, will it take
to fill the pool 6 inches from the top? [1 ft^3 = 7.48 gallons]

HOW TO CONVERT YOUR RAW SCORE TO YOUR GEOMETRY REGENTS EXAMINATION SCORE

The conversion chart below can be used to determine your final score on the August 2017 Regents Exam in Geometry. To find your final exam score, locate in the "Raw Score" column the total number of points you scored out of a possible 86. Then locate in the adjacent column to the right the scale score that corresponds to your raw score. The scale score is your final score.

Raw Score	Scale Score	Performance Level	Raw Score	Scale Score	Performance Level	Raw Score	Scale Score	Performance Level
86	100	5	57	79	3	28	59	2
85	99	5	56	79	3	27	58	2
84	97	5	55	78	3	26	57	2
83	96	5	54	78	3	25	55	2
82	95	5	53	77	3	24	54	1
81	94	5	52	77	3	23	53	1
80	93	5	51	76	3	22	51	1
79	92	5	50	76	3	21	50	1
78	91	5	49	75	3	20	48	1
77	90	5	48	75	3	19	47	1
76	90	5	47	74	3	18	45	1
75	89	5	46	74	3	17	43	1
74	88	5	45	73	3	16	42	1
73	87	5	44	73	3	15	40	1
72	87	5	43	72	3	14	38	1
71	86	5	42	72	3	13	36	1
70	86	5	41	71	3	12	34	1
69	86	5	40	70	3	11	32	1
68	85	5	39	70	3	10	30	1
67	84	4	38	69	3	9	28	1
66	83	4	37	68	3	8	25	1
65	83	4	36	67	3	7	23	1
64	82	4	35	66	3	6	20	1
63	82	4	34	66	3	5	18	1
62	82	4	33	65	3	4	15	1
61	81	4	32	64	2	3	12	1
60	81	4	31	63	2	2	8	1
59	80	4	30	61	2	1	5	1
58	80	3	29	60	2	0	0	1

Answers
August 2017
Geometry (Common Core)

Answer Key

PART I

1. 2	**5.** 4	**9.** 4	**13.** 3	**17.** 1	**21.** 4
2. 4	**6.** 3	**10.** 1	**14.** 3	**18.** 2	**22.** 4
3. 3	**7.** 4	**11.** 4	**15.** 1	**19.** 1	**23.** 2
4. 1	**8.** 4	**12.** 2	**16.** 4	**20.** 1	**24.** 4

PART II

25. The heights and cross-sectional areas are equal, so the volumes must be equal.

26. 68

27. A rotation of 180° about the origin followed by a translation of 1 unit up and 1 unit left

28. See detailed solution for the construction.

29. $\dfrac{5}{2}$

30. Rigid motions preserve angle measure and length, so the triangles are congruent and $\overline{BC} \cong \overline{YZ}$ by CPCTC.

31. Center $(3, -4)$ radius $= 9$

PART III

32. Slopes of $\overline{AB}$ and $\overline{PR}$ are both $\dfrac{2}{3}$. $AB = \sqrt{13}$ and $\overline{PR} = 2\sqrt{13}$.

33. Prove $\triangle ABC$ is similar to $\triangle ECA$, then form a proportion from corresponding sides.

34. 2402.2 cm^2

PART IV

35. See the detailed solution for the proof.

36. Depth $= 8.5$ ft, volume $= 3{,}752$ ft^3, time $= 41$ hours

In **PARTS II–IV,** you are required to show how you arrived at your answers. For sample methods of solutions, see the *Answers Explained* section.

Answers and Explanations

PART I

1. The only choice that does not have a triangular cross-section is the cylinder. The triangular cross-sections of the other three choices are shown in the accompanying figure.

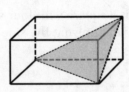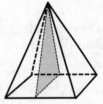

The correct choice is **(2)**.

2. Reflections are rigid motions that always result in a congruent image. Dilations maintain the original shape but may change the size if the scale factor is different from 1. A dilation with a scale factor of $\frac{3}{2}$ will result in an image larger than the original.

The correct choice is **(4)**.

3. Use the distance formula and the given coordinates $(-1, 5)$ and $(-3, 1)$ to find the length of side $\overline{RS}$.

$$
\begin{aligned}
d &= \sqrt{\left(x_2 - x_1\right)^2 + \left(y_2 - y_1\right)^2} \\
&= \sqrt{\left(-3(-1)\right)^2 + (1 - 5)^2} \\
&= \sqrt{(-2)^2 + (-4)^2} \\
&= \sqrt{20}
\end{aligned}
$$

Since we know the figure is a square with four sides of equal length, we can calculate the perimeter by multiplying the side length by 4. The perimeter is $4\sqrt{20}$.

The correct choice is **(3)**.

4. The arcs between parallel chords are congruent, therefore $m\overarc{AC} = m\overarc{BD}$ and each can be represented by x. The diameter $\overline{AOB}$ intercepts a semi-circle whose arcs sum to $180°$.

$$
\begin{aligned}
m\overarc{AC} + m\overarc{CD} + m\overarc{BD} &= 180 \\
x + 130° + x &= 180° \\
2x + 130° &= 180° \\
2x &= 50° \\
x &= 25°
\end{aligned}
$$

The correct choice is **(1)**.

5. $\angle A \cong \angle D$ because they are alternate interior angles formed by the parallel segments $\overline{AB}$ and $\overline{DE}$. Also, the vertical angles $\angle ACB$ and $\angle DCE$ are congruent. Therefore, $\triangle ACB$ is similar to $\triangle DCE$ by the AA theorem and corresponding sides lengths are proportional.

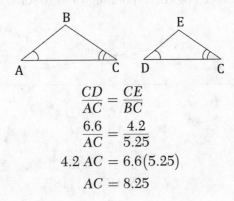

$$\frac{CD}{AC} = \frac{CE}{BC}$$
$$\frac{6.6}{AC} = \frac{4.2}{5.25}$$
$$4.2\,AC = 6.6(5.25)$$
$$AC = 8.25$$

The correct choice is **(4)**.

6. A reflection over the x-axis would map the top and bottom halves of the rectangle onto each other. The line $x = 4$ is a vertical line through the center of the rectangle, and it would map the left and right halves onto each other. The point $(4, 0)$ is in the center of the rectangle. A rotation of $180°$ about that center point would also result in the rectangle mapping onto itself. The only choice that does not map the rectangle onto itself is a $180°$ rotation about the origin.

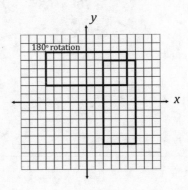

The correct choice is **(3)**.

7. Sketch the two triangles separately as shown to help identify congruent angles and corresponding sides. The parallel segments $\overline{BE}$ and $\overline{CD}$ form congruent alternate interior angles $\angle ABE$ and $\angle ACD$. $\angle A$ is a shared angle and is congruent to itself. Therefore, the two triangles are similar by the AA postulate. Write a proportion using corresponding sides to solve for the length AE.

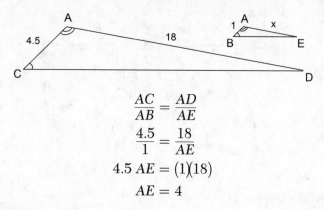

$$\frac{AC}{AB} = \frac{AD}{AE}$$
$$\frac{4.5}{1} = \frac{18}{AE}$$
$$4.5\,AE = (1)(18)$$
$$AE = 4$$

The correct choice is **(4)**.

8. Opposite angles of a parallelogram are congruent, so $m\angle R = m\angle C = 70°$. $\angle KSO$ is an exterior angle of $\triangle RSO$. From the exterior angle theorem we know the measure of the exterior angle is equal to the sum of the two further interior angles.

$$m\angle KSO = m\angle R + m\angle ROS$$
$$= 70° + 65°$$
$$= 135°$$

The correct choice is **(4)**.

9. Check if the two pairs of corresponding sides are proportional. Form ratios using the two shorter sides and the two longer sides. If it is a true proportion the cross products will be equal.

$$\frac{AR}{GR} = \frac{TR}{SR}$$

$$\frac{15}{36} = \frac{18}{45}$$

$$15 \cdot 45 = 18 \cdot 36$$

$$675 \neq 638$$

The cross products are not equal, so the two pairs of sides are not proportional. The triangles are not similar.

The correct choice is (**4**).

10. Dilating a line through the origin will multiply the y-intercept by the scale factor, but leave the slope unchanged. We need to find which choice has the same slope as the given line. Rewrite the given equation in $y = mx + b$ form to identify the slope.

$$4y = 3x + 7$$

$$y = \frac{3}{4}x + \frac{7}{4}$$

The y-intercept is $\frac{7}{4}$ and the slope is $\frac{3}{4}$. Now check the slope of each choice by writing in $y = mx + b$ form. For choice (1) we have

$$3x - 4y = 9$$

$$-4y = -3x + 9$$

$$y = \frac{3}{4}x - \frac{9}{4}$$

Equation (1) also has a slope of $\frac{3}{4}$, so this must be the dilation.

The correct choice is (**1**).

11. The isosceles triangle theorem states that if $\overline{AB} \cong \overline{BC}$ then the opposite angles must also be congruent. Therefore, $\angle BAC \cong \angle BCA$ and we can represent both angles with the variable y (not x since that variable is already used for another angle!). Apply the angle sum theorem in $\triangle ABC$.

$$m\angle A + m\angle B + m\angle ACB = 180°$$
$$y + 62° + y = 180°$$
$$2y + 62° = 180°$$
$$2y = 180°$$
$$y = 59°$$

$m\angle A$ and $m\angle ACB$ both measure $59°$. The angle labeled x is an exterior angle, so its measure is equal to the sum of the two furthest interior angles.

$$x = m\angle A + m\angle B$$
$$x = 59° + 62°$$
$$x = 121°$$

The correct choice is **(4)**.

12. The lengths of the secant and tangent are related by the formula

$$(\text{secant})^2 = (\text{outside part of tangent}) \cdot (\text{entire tangent})$$

$$PA^2 = PB(PB + PC)$$
$$PA^2 = 3(3 + 15)$$
$$PA^2 = 3(18)$$
$$PA^2 = 54$$
$$PA = \sqrt{54}$$

Since this is not one of the choices we simplify the radical. Look for a factor of 54 that is a perfect square. The desired factors are 9 and 6, where 9 is the perfect square.

$$\sqrt{54} = \sqrt{9 \cdot 6}$$
$$= 3\sqrt{6}$$

The correct choice is (2).

As an alternative to simplifying the radical, you can type the radical into your calculator to find the decimal approximation 7.34846. Check each of the choices to find the one with the matching decimal approximation.

13. Rotating a rectangle around a side will form a cylinder. The length of the side it is rotated around is the height and the other side gives the radius as shown in figures (a) and (b) below. The situation is similar if the rectangle is rotated about a line of symmetry, except that one of the sides gives the diameter instead of the radius as shown in figure (c) below.

(a) rotation about the short side　　(b) rotation about the long side　　(c) rotation about the line of symmetry

Calculate the volume for each choice using the volume formula $V = \pi R^2 h$, which is provided on your reference sheet.

Choice 1:
$h = 10,\ R = 6$

$V = \pi R^2 h$

$\quad = \pi \cdot 6^2 \cdot 10$

$\quad = 360\pi$

Choice 3:
$h = 6,\ R = 5$

$V = \pi R^2 h$

$\quad = \pi \cdot 5^2 \cdot 6$

$\quad = 150\pi$

Choice 2:
$h = 6,\ R = 10$

$V = \pi R^2 h$

$\quad = \pi \cdot 10^2 \cdot 6$

$\quad = 600\pi$

Choice 4:
$h = 10,\ R = 3$

$V = \pi R^2 h$

$\quad = \pi \cdot 3^2 \cdot 10$

$\quad = 90\pi$

14. A parallelogram whose diagonals are perpendicular is a rhombus. $\overline{AC} \perp \overline{BD}$ represents the two perpendicular diagonals.

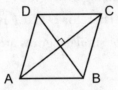

The correct choice is **(3)**.

The other possible rhombus properties that could be used are consecutive sides congruent or diagonals bisect the vertex angles. However, none of the choices represents these properties.

15. Find the unknown angle by setting up a trig ratio using an inverse trig function. Relative to the angle x, the opposite side has length 1 and the adjacent side has length 12. These suggest using the tangent.

$$\tan(x) = \frac{\text{opposite}}{\text{adjacent}}$$

$$= \frac{1}{12}$$

$$x = \tan^{-1}\left(\frac{1}{12}\right)$$

$$= 4.76$$

The correct choice is **(1)**.

16. The three sides of $\triangle DEF$ are all midsegments because they are formed by joining midpoints of the sides of $\triangle ABC$. A midsegment has a length equal to $\frac{1}{2}$ of the opposite side, making every side in $\triangle DEF$ half the length of a corresponding side in $\triangle ABC$. Therefore, $\triangle DEF$ is the image after a dilation with a scale factor of $\frac{1}{2}$. Areas of similar figures are proportional to the scale factor squared.

$$\text{Ratio of areas} = \left(\frac{1}{2}\right)^2$$
$$= \frac{1}{4}$$

The correct choice is **(4)**.

17. Find the x- and y-coordinates of P with the formula

$$\text{ratio} = \frac{x - x_1}{x_2 - x} \qquad \text{ratio} = \frac{y - y_1}{y_2 - y}$$

Using the coordinates $A(-8, -2)$ and $B(16, 6)$, we have $x_1 = -8$, $y_1 = -2$, $x_2 = 16$, and $y_2 = 6$. The ratio is $\frac{3}{5}$.

Solve for the x-coordinate:

$$\text{ratio} = \frac{x - x_1}{x_2 - x}$$
$$\frac{3}{5} = \frac{x - (-8)}{16 - x}$$
$$5(x + 8) = 3(16 - x) \qquad \text{cross-multiply}$$
$$5x + 40 = 48 - 3x$$
$$8x = 8$$
$$x = 1$$

Solve for the y-coordinate:

$$\text{ratio} = \frac{y - y_1}{y_2 - y}$$

$$\frac{3}{5} = \frac{y - (-2)}{6 - y}$$

$$5(y + 2) = 3(6 - y) \qquad \text{cross-multiply}$$

$$5y + 10 = 18 - 3y$$

$$8y = 8$$

$$y = 1$$

The coordinates of P are $(1, 1)$.

The correct choice is (**1**).

18. One possible approach for this problem is to calculate the measures of $\angle K$ and $\angle M$ using the given lengths and assuming $\angle KLM$ is a right angle. If the sum of $\angle K$ and $\angle M$ is $90°$, then our assumption was correct and $\angle KLM$ must be a right angle. Using choice (2), we find

$$\cos M = \left(\frac{9}{12}\right) \text{ in } \triangle KLM \qquad \sin K = \frac{12}{16} \text{ in } \triangle KLM$$

$$M = \cos^{-1}\frac{9}{12} \qquad\qquad K = \sin^{-1}\frac{12}{16}$$

$$= 41.4° \qquad\qquad\qquad = 48.6°$$

Since $41.4° + 48.6° = 90°$, we know the third angle is $90°$ as well.

The correct choice is (**2**).

19. Apply a trig ratio to find BC. Relative to $\angle A$, $\overline{AC}$ is the hypotenuse and $\overline{BC}$ is the opposite. We use the sine ratio.

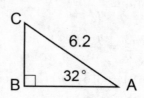

$$\sin(32) = \frac{\text{opp}}{\text{hyp}}$$
$$\sin(32) = \frac{BC}{6.2}$$
$$BC = 6.2\sin(32)$$
$$= 3.3$$

The correct choice is (**1**).

20. The population density, area, and population are related by the formula

$$\text{population density} = \frac{\text{population}}{\text{area}}$$

Since we need to find the area of four states, it will be easiest to solve for area in terms of the other variables.

$$\text{population density} = \frac{\text{population}}{\text{area}}$$
$$\text{area} \cdot \text{population density} = \text{population}$$
$$\text{area} = \frac{\text{population}}{\text{population density}}$$

Florida

$$\text{area} = \frac{18,801,310}{350.6}$$

$$= 53,626 \text{ miles}^2$$

New York

$$\text{area} = \frac{19,378,102}{411.2}$$

$$= 47,126 \text{ miles}^2$$

Illinois

$$\text{area} = \frac{12,830,632}{231.1}$$

$$= 55,520$$

Pennsylvania

$$\text{area} = \frac{12,702,379}{283.9}$$

$$= 44,742$$

The order of states from largest area to smallest is Illinois, Florida, New York, and Pennsylvania.

The correct choice is (**1**).

21. The co-function relationship states that if a sine and cosine are equal, then the angles must sum to 90°. Set the sum of the two angles to 90° and then solve for x.

$$40 - x + 3x = 90$$
$$40 + 2x = 90$$
$$2x = 50$$
$$x = 25$$

The correct choice is (**4**).

22. A rotation of any multiple of $\frac{360}{n}$, where n is the number of sides, will map a regular polygon onto itself. A decagon has ten sides, so the angle must be a multiple of $\frac{360}{10}$ or 36°. Divide each choice by 36°. The one with an integer quotient is a multiple of 36°.

$$\frac{10}{36} = 0.27 \qquad \frac{225}{36} = 6.25$$
$$\frac{150}{36} = 4.17 \qquad \frac{252}{36} = 7$$

A rotation of 252° will map a decagon onto itself.

The correct choice is **(4)**.

23. The radius of the circle is half the diameter, or 16. Apply the formula for the area of a sector to find the angle.

$$A_{sector} = \frac{1}{2}R^2, \text{where } \theta \text{ is the central angle measured in radians}$$
$$\frac{512\pi}{3} = \frac{1}{2} \cdot 16^2$$
$$\frac{1024\pi}{3} = 16^2$$
$$= \frac{1024}{3.16^2}$$
$$= \frac{4\pi}{3}$$

The correct choice is **(2)**.

24. The perpendicular bisector passes through the midpoint of the given segment, and has a slope equal to the negative reciprocal of the segment's slope. Find the midpoint using the endpoints $(-2, 4)$ and $(2, -4)$ and the midpoint formula.

$$\text{midpoint} = \left(\frac{x_1 + x_2}{2}, \frac{y_1 + y_2}{2} \right)$$

$$= \left(\frac{-2 + 2}{2}, \frac{4 + (-4)}{2} \right)$$

$$= (0, 0)$$

Next find the slope of the segment.

$$\text{slope} = \frac{y_2 - y_1}{x_2 - x_1}$$

$$= \frac{-4 - 4}{2 - (-2)}$$

$$= \frac{-8}{4}$$

$$= -2$$

The negative reciprocal of the slope is $\frac{1}{2}$. Using the point–slope form of a line with a slope of $\frac{1}{2}$ and the point $(0, 0)$ we have

$$y - y_1 = m(x - x_1)$$

$$y - 0 = \frac{1}{2}(x - 0)$$

$$y = \frac{1}{2}x$$

$$y - \frac{1}{2}x = 0$$

$$2y - x = 0 \qquad \text{multiply both sides by 2 to eliminate the fraction}$$

The correct choice is **(4)**.

PART II

25. The volumes are the same. Each cylinder has a radius of 5 meters, so the cross-sectional areas are equal. Also they each have a height of 11.5 meters. Cavalieri's principle states that if two solids have equal cross-sectional areas and heights, then their volumes are equal. *Note that calculating the volume of each cylinder would earn only 1 of the 2 possible credits.*

26. The diagonals of a rhombus, like all parallelograms, bisect each other. Calculate the lengths of $\overline{TS}$ and $\overline{TR}$ by dividing each diagonal by 2. $TS = 15$ and $TR = 8$. The diagonals of rhombuses are also perpendicular, so $\triangle STR$ is a right triangle. SR can be found using the Pythagorean theorem.

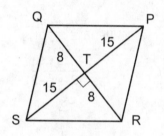

$$a^2 + b^2 = c^2$$
$$15^2 + 8^2 = SR^2$$
$$289 = SR^2$$
$$SR = 17$$

All four sides of a rhombus are congruent, so the perimeter is equal to four times the side length.

$$\text{perimeter} = 4(17)$$
$$= 68$$

27. A rotation of 180° about the origin followed by a translation of 1 unit up and 1 unit left will map *MATH* to *M″A″T″H″*.

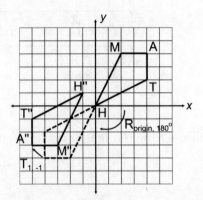

28. Mark point *A* anywhere on the circle. Using the same radius as the circle and the point of the compass at *A*, make an arc intersecting the circle at *B*. With the same compass opening, place the point of the compass at *B* and make an arc at *C*. Continue making arcs intersecting at *D*, *E*, and *F*. *ABCDEF* is a regular hexagon.

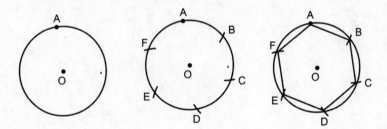

29. The length of $\overline{AB}$ can be calculated using the distance formula and coordinates $A(2, 3)$ and $B(5, -1)$.

$$
\begin{aligned}
d &= \sqrt{\left(x_2 - x_1\right)^2 + \left(y_2 - y_1\right)^2} \\
&= \sqrt{(5 - 2)^2 + (-1 - 3)^2} \\
&= \sqrt{25} \\
&= 5
\end{aligned}
$$

A dilation of $\frac{1}{2}$ will multiply the length of the segment by $\frac{1}{2}$, so the length of the image is $\frac{5}{2}$.

30. Rigid motions preserve angle measure and length, so $\angle A \cong \angle X$, $\angle C \cong \angle Z$, and $\overline{AC} \cong \overline{XZ}$. The triangles are therefore congruent by the ASA postulate. $\overline{BC} \cong \overline{YZ}$ because corresponding parts of congruent triangles are congruent.

31. Use completing the square to rewrite the equation in the form $(x - h)^2 + (y - k)^2 = r^2$. First group the x-terms and the y-terms on the left, and move the constant term to the right.

$$x^2 + y^2 - 6x = 56 - 8y$$
$$x^2 - 6x + y^2 + 8y = 56$$

Next find the constant terms needed to complete the square:

constant for the x − terms	constant for the y − terms
$\left(\dfrac{1}{2} \text{ coefficient of } x\right)^2$	$\left(\dfrac{1}{2} \text{ coefficient of } y\right)^2$
$= \left(\dfrac{1}{2} \cdot (-6)\right)^2$	$= \left(\dfrac{1}{2} \cdot 8\right)^2$
$= (-3)^2$	$= (4)^2$
$= 9$	$= 16$

Add the required constants to each side of the equation and factor the left-hand side.

$x^2 - 6x + 9 + y^2 + 8y + 16 = 56 + 9 + 16$ add 9 and 16 to each side

$x^2 - 6x + 9 + y^2 + 8y + 16 = 81$ simplify

$\left(x - 3\right)^2 + \left(y + 4\right)^2 = 9$ factor

The values of h and k are 3 and -4. Be careful with the signs because h and k are the values subtracted from x and y. The center is located at $(3, -4)$. Since $r^2 = 9$, the radius is equal to 3.

center: (**3**, **−4**) radius: (**3**).

PART III

32. Find the coordinates of the midpoints A and B using the midpoint formula.

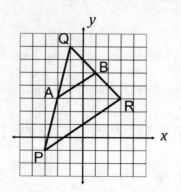

midpoint of $P(-3, -1)\,Q(-1, 7)$ midpoint of $Q(-1, 7)\,R(3, 3)$

$$\text{midpoint} = \left(\frac{x_1 + x_2}{2}, \frac{y_1 + y_2}{2}\right) \qquad \text{midpoint} = \left(\frac{x_1 + x_2}{2}, \frac{y_1 + y_2}{2}\right)$$

$$= \left(\frac{-3 + (-1)}{2}, \frac{-1 + 7}{2}\right) \qquad\qquad = \left(\frac{-1 + 3}{2}, \frac{7 + 3}{2}\right)$$

$$= (-2, 3) \qquad\qquad\qquad\qquad\qquad = (1, 5)$$

Next find the length and slope of segments $\overline{AB}$ and $\overline{PR}$.

$$\text{slope of } \overline{AB} = \frac{y_2 - y_1}{x_2 - x_1} \qquad \text{slope of } \overline{PR} = \frac{y_2 - y_1}{x_2 - x_1}$$

$$= \frac{5 - 3}{1 - (-2)} \qquad\qquad = \frac{3 - (-1)}{3 - (-3)}$$

$$= \frac{2}{3} \qquad\qquad\qquad = \frac{4}{6}$$

$$\qquad\qquad\qquad\qquad\qquad = \frac{2}{3}$$

$$AB = \sqrt{\left(x_2 - x_1\right)^2 + \left(y_2 - y_1\right)^2} \qquad PR = \sqrt{\left(x_2 - x_1\right)^2 + \left(y_2 - y_1\right)^2}$$

$$= \sqrt{\left(1 - (-2)\right)^2 + (5 - 3)^2} \qquad\qquad = \sqrt{\left(3 - (-3)\right)^2 + \left(3 - (-1)\right)^2}$$

$$= \sqrt{9 + 4} \qquad\qquad\qquad\qquad = \sqrt{36 + 16}$$

$$= \sqrt{13} \qquad\qquad\qquad\qquad = \sqrt{52}$$

$$\qquad\qquad\qquad\qquad\qquad = \sqrt{4 \cdot 13}$$

$$\qquad\qquad\qquad\qquad\qquad = 2\sqrt{13}$$

$\overline{AB}$ is parallel to $\overline{PR}$ because they have the same slope. $\overline{AB}$ is half the length of $\overline{PR}$ because $\sqrt{13}$ is one-half of $2\sqrt{13}$.

33. To prove two of the pairs of sides are proportional we can prove $\triangle ABC$ is similar to $\triangle ECA$. $\angle B$ is a right angle because it is an inscribed angle that intercepts a diameter. $\angle ECA$ is a right angle because it is formed by a tangent and radius. Therefore, $\angle B \cong \angle ECA$. We also have $\angle BCA \cong \angle ECA$ because they are the alternate interior angles formed by the given parallel lines. Since two pairs of angles are congruent, $\triangle ABC$ is similar to $\triangle ECA$ by the AA postulate. $\dfrac{BC}{CA} = \dfrac{AB}{EC}$ because the corresponding parts of similar triangles are proportional.

34. Let s represent the side length of the poster. The 58 centimeter diagonal forms a right isosceles triangle with side length s. Use the Pythagorean theorem to solve for s.

$$a^2 + b^2 = c^2$$
$$s^2 + s^2 = 58^2$$
$$2s^2 = 3,364$$
$$s^2 = 1,682$$
$$s^2 = \sqrt{1,682}$$
$$s = 41.012$$

The length of the frame L can be found by adding 4 centimeters to each side of the poster.

$$L = 41.012 + 4 + 4$$
$$= \mathbf{49.012}$$

The area of the square frame is found using the area formula for a square.

$$\text{Area} = L^2$$
$$= 49.012^2$$
$$= 2402.2195$$
$$= \mathbf{2402.2 \ cm^2}$$

PART IV

35. The strategy for this proof is to show △*DEC* is isosceles. The congruent legs $\overline{DE}$ and $\overline{EC}$ can be used to prove △*ADE* and △*BCE* are congruent. CPCTC can then be used to show △*AEB* is isosceles.

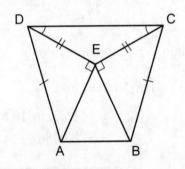

Statement	Reason
1. ∠*CDE* ≅ ∠*DEC*	1. Given
2. $\overline{DE}$ ≅ $\overline{EC}$	2. Sides of a triangle opposite congruent angles are congruent
3. $\overline{AE}$ and $\overline{DE}$, $\overline{BE}$ and $\overline{CE}$	3. Given
4. *AED* and *BEC* are right angles	4. Perpendicular lines form right angles
5. △*AED* and △*BEC* are right triangles	5. A triangle with a right angle is a right triangle
6. Isosceles trapezoid *ABCD*	6. Trigonometric Expressions and Equations
7. $\overline{AD}$ ≅ $\overline{BC}$	7. Legs of an isosceles trapezoid are congruent
8. △*AED* ≅ △*BEC*	8. HL
9. $\overline{AE}$ ≅ $\overline{BE}$	9. CPCTC
10. △*AEB* is isosceles	10. A triangle with two congruent sides is isosceles

36. The pool is composed of three rectangular prisms and a triangular prism as shown in the figure below. The dimensions x and y are needed in the triangular prism in order to calculate its base area. The dimension y is also needed to determine the height of prism D.

$$x = 35 - 9 - 12.5$$
$$= 13.5$$

Apply the tangent ratio to find the dimension x.

$$\tan(16.5) = \frac{y}{13.5}$$
$$y = 13.5\tan(16.5)$$
$$= 3.9988$$

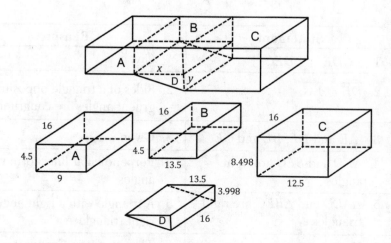

Now calculate the volume of each prism.

Prism A

$V = Bh$

$= (9 \cdot 16)4.5$

$= 648 \text{ ft}^3$

Prism C

$V = Bh$

$= \left(\frac{1}{2}13.5 \cdot 3.9988\right)16$

$= 431.8704 \text{ ft}^3$

Prism B

$V = Bh$

$= (13.5 \cdot 16)4.5$

$= 972 \text{ ft}^3$

Prism D

$V = Bh$

$= (12.5 \cdot 16)(4.5 + 3.9988)$

$= 1699.76 \text{ ft}^3$

The total volume of the pool is

$$648 + 972 + 431.8704 + 1699.76 = 3751.6304.$$

Rounded to the nearest cubic foot, the volume is 3752 ft^3.

To find how long it will take to fill the pool 6 inches from the top we need to subtract the volume of a thin 6-inch-high rectangular prism from the total volume. This volume is taken from the very top of the pool, so its dimensions are 16 feet × 35 feet × 6 inches. Convert the 6 inches to feet before calculating the volume.

$$6 \text{ inches} \cdot \frac{1 \text{ foot}}{12 \text{ inches}} = 0.5 \text{ foot}$$

$$\text{The volume of the top layer} = B\,h$$

$$= (16.35)\,0.5$$

$$= 280 \text{ ft}^3$$

$$\text{The volume to be filled} = 3752 \text{ ft}^3 - 280 \text{ ft}^3$$

$$= 3472 \text{ ft}^3$$

Next, convert the volume from cubic feet to gallons using the conversion ratio provided in the problem.

$$\text{Number of gallons} = 3472 \text{ ft}^3 \cdot \frac{7.48 \text{ gallons}}{\text{ft}^3}$$
$$= 25970.56 \text{ gallons}$$

Now calculate the time required to fill the pool using the fill rate $\frac{10.5 \text{ gallons}}{\text{minute}}$. Since the gallons per minute ratio has gallons in the numerator, we need to *divide* by this ratio.

$$\text{time} = 25970.56 \text{ gallons} \div \frac{10.5 \text{ gallons}}{\text{minute}}$$
$$= 25970.56 \cdot \frac{1 \text{ minute}}{10.5 \text{ gallons}}$$
$$= 2473.38 \text{ minutes}$$

Finally, convert the time in minutes to time in hours using the ratio $\frac{1 \text{ hour}}{60 \text{ minutes}}$

$$\text{time} = 2473.38 \text{ minutes} \cdot \frac{1 \text{ hour}}{60 \text{ minutes}}$$
$$= 41.22 \text{ hours}$$

Rounded to the nearest hour, it will take 41 hours to fill the pool.

Examination
June 2018
Geometry

GEOMETRY REFERENCE SHEET

1 inch = 2.54 centimeters	1 ton = 2000 pounds
1 meter = 39.37 inches	1 cup = 8 fluid ounces
1 mile = 5280 feet	1 pint = 2 cups
1 mile = 1760 yards	1 quart = 2 pints
1 mile = 1.609 kilometers	1 gallon = 4 quarts
1 kilometer = 0.62 mile	1 gallon = 3.785 liters
1 pound = 16 ounces	1 liter = 0.264 gallon
1 pound = 0.454 kilogram	1 liter = 1000 cubic centimeters
1 kilogram = 2.2 pounds	

Triangle	$A = \dfrac{1}{2}bh$
Parallelogram	$A = bh$
Circle	$A = \pi r^2$
Circle	$C = \pi d$ or $C = 2\pi r$

General Prisms	$V = Bh$
Cylinder	$V = \pi r^2 h$
Sphere	$V = \frac{4}{3}\pi r^3$
Cone	$V = \frac{1}{3}\pi r^2 h$
Pyramid	$V = \frac{1}{3}Bh$
Pythagorean Theorem	$a^2 + b^2 = c^2$
Quadratic Formula	$x = \dfrac{-b \pm \sqrt{b^2 - 4ac}}{2a}$
Arithmetic Sequence	$a_n = a_1 + (n-1)d$
Geometric Sequence	$a_n = a_1 r^{n-1}$
Geometric Series	$S_n = \dfrac{a_1 - a_1 r^n}{1 - r}$ where $r \neq 1$
Radians	$1 \text{ radian} = \dfrac{180}{\pi} \text{ degrees}$
Degrees	$1 \text{ degree} = \dfrac{\pi}{180} \text{ radians}$
Exponential Growth/Decay	$A = A_0 e^{k(t-t_0)} + B_0$

180 Minutes–35 Questions

PART I

Answer all 24 questions in this part. Each correct answer will receive 2 credits. No partial credit will be allowed. For each statement or question, write in the space provided the numeral preceding the word or expression that best completes the statement or answers the question.   [48 credits]

1 After a counterclockwise rotation about point X, scalene triangle ABC maps onto $\triangle RST$, as shown in the diagram below.

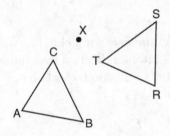

Which statement must be true?

(1) $\angle A \cong \angle R$

(2) $\angle A \cong \angle S$

(3) $\overline{CB} \cong \overline{TR}$

(4) $\overline{CA} \cong \overline{TS}$ 1 _____

2 In the diagram below, $\overline{AB} \parallel \overline{DEF}$, $\overline{AE}$ and $\overline{BD}$ intersect at C, $m\angle B = 43°$, and $m\angle CEF = 152°$.

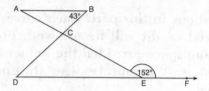

Which statement is true?

(1) $m\angle D = 28°$ (3) $m\angle ACD = 71°$

(2) $m\angle A = 43°$ (4) $m\angle BCE = 109°$ 2 _____

3 In the diagram below, line m is parallel to line n. Figure 2 is the image of Figure 1 after a reflection over line m. Figure 3 is the image of Figure 2 after a reflection over line n.

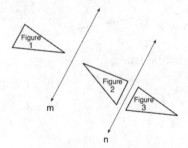

Which single transformation would carry Figure 1 onto Figure 3?

(1) a dilation   (3) a reflection

(2) a rotation   (4) a translation   3 _____

4 In the diagram below, $\overline{AF}$ and $\overline{DB}$ intersect at C, and $\overline{AD}$ and $\overline{FBE}$ are drawn such that m$\angle D = 65°$, m$\angle CBE = 115°$, $DC = 7.2$, $AC = 9.6$, and $FC = 21.6$.

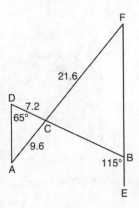

What is the length of $\overline{CB}$?

(1) 3.2   (3) 16.2

(2) 4.8   (4) 19.2   4 _____

5 Given square $RSTV$, where $RS = 9$ cm. If square $RSTV$ is dilated by a scale factor of 3 about a given center, what is the perimeter, in centimeters, of the image of $RSTV$ after the dilation?

(1) 12   (3) 36

(2) 27   (4) 108   5 _____

6 In right triangle ABC, hypotenuse $\overline{AB}$ has a length of 26 cm, and side $\overline{BC}$ has a length of 17.6 cm. What is the measure of angle B, to the *nearest degree*?

(1) 48°   (3) 43°

(2) 47°   (4) 34°   6 _____

7 The greenhouse pictured below can be modeled as a rectangular prism with a half-cylinder on top. The rectangular prism is 20 feet wide, 12 feet high, and 45 feet long. The half-cylinder has a diameter of 20 feet.

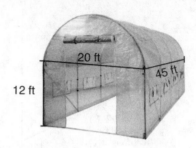

To the *nearest cubic foot*, what is the volume of the greenhouse?

(1) 17,869   (3) 39,074

(2) 24,937   (4) 67,349   7 _____

8 In a right triangle, the acute angles have the relationship $\sin(2x + 4) = \cos(46)$.

What is the value of x?

(1) 20   (3) 24

(2) 21   (4) 25   8 _____

9 In the diagram below, $\overline{AB} \parallel \overline{DFC}$, $\overline{EDA} \parallel \overline{CBG}$, and $\overline{EFB}$ and $\overline{AG}$ are drawn.

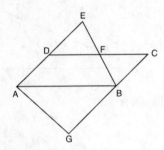

Which statement is always true?

(1) $\triangle DEF \cong \triangle CBF$ (3) $\triangle BAG \sim \triangle AEB$

(2) $\triangle BAG \cong \triangle BAE$ (4) $\triangle DEF \sim \triangle AEB$ 9 _____

10 The base of a pyramid is a rectangle with a width of 4.6 cm and a length of 9 cm. What is the height, in centimeters, of the pyramid if its volume is 82.8 cm^3?

(1) 6   (3) 9

(2) 2   (4) 18   10 _____

11 In the diagram below of right triangle AED, $\overline{BC} \parallel \overline{DE}$.

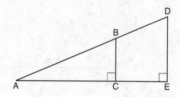

Which statement is always true?

(1) $\dfrac{AC}{BC} = \dfrac{DE}{AE}$ (3) $\dfrac{AC}{CE} = \dfrac{BC}{DE}$

(2) $\dfrac{AB}{AD} = \dfrac{BC}{DE}$ (4) $\dfrac{DE}{BC} = \dfrac{DB}{AB}$ 11 _____

12 What is an equation of the line that passes through the point $(6, 8)$ and is perpendicular to a line with equation $y = \dfrac{3}{2}x + 5$?

(1) $y - 8 = \dfrac{3}{2}(x - 6)$ (3) $y + 8 = \dfrac{3}{2}(x + 6)$

(2) $y - 8 = -\dfrac{2}{3}(x - 6)$ (4) $y + 8 = -\dfrac{2}{3}(x + 6)$ 12 _____

13 The diagram below shows parallelogram $ABCD$ with diagonals $\overline{AC}$ and $\overline{BD}$ intersecting at E.

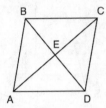

What additional information is sufficient to prove that parallelogram $ABCD$ is also a rhombus?

(1) $\overline{BD}$ bisects $\overline{AC}$.

(2) $\overline{AB}$ is parallel to $\overline{CD}$.

(3) $\overline{AC}$ is congruent to $\overline{BD}$.

(4) $\overline{AC}$ is perpendicular to $\overline{BD}$.   13 _____

14 Directed line segment DE has endpoints $D(-4, -2)$ and $E(1, 8)$. Point F divides $\overline{DE}$ such that $DF : FE$ is $2 : 3$. What are the coordinates of F?

(1) $(-3, 0)$ (3) $(-1, 4)$

(2) $(-2, 2)$ (4) $(2, 4)$ 14 _____

15 Triangle *DAN* is graphed on the set of axes below. The vertices of △*DAN* have coordinates *D*(−6, −1), *A*(6, 3), and *N*(−3, 10).

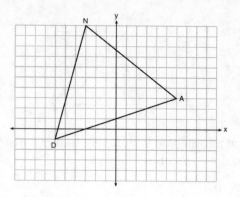

What is the area of △*DAN*?

(1) 60

(2) 120

(3) 20√13

(4) 40√13

15 _____

16 Triangle ABC, with vertices at $A(0, 0)$, $B(3, 5)$, and $C(0, 5)$, is graphed on the set of axes shown below.

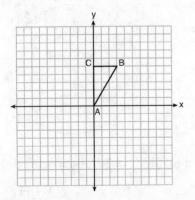

Which figure is formed when $\triangle ABC$ is rotated continuously about $\overline{BC}$?

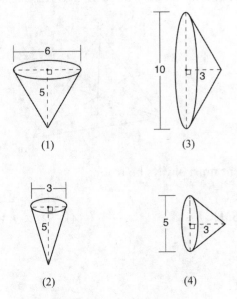

(1)

(2)

(3)

(4)

16 ____

17 In the diagram below of circle O, chords $\overline{AB}$ and $\overline{CD}$ intersect at E.

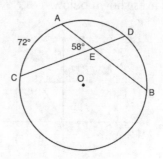

If m $\overset{\frown}{AC} = 72°$ and m$\angle AEC = 58°$, how many degrees are in m $\overset{\frown}{DB}$?

(1) 108°   (3) 44°

(2) 65°   (4) 14°   17 _____

18 In triangle SRK below, medians $\overline{SC}$, $\overline{KE}$, and $\overline{RL}$ intersect at M.

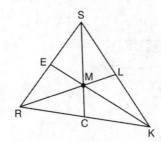

Which statement must always be true?

(1) $3(MC) = SC$ (3) $RM = 2MC$

(2) $MC = \frac{1}{3}(SM)$ (4) $SM = KM$ 18 _____

19 The regular polygon below is rotated about its center.

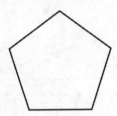

Which angle of rotation will carry the figure onto itself?

(1) 60°   (3) 216°

(2) 108°   (4) 540°   19 _____

20 What is an equation of circle O shown in the graph below?

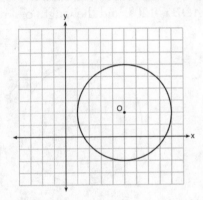

(1) $x^2 + 10x + y^2 + 4y = -13$

(2) $x^2 - 10x + y^2 - 4y = -13$

(3) $x^2 + 10x + y^2 + 4y = -25$

(4) $x^2 - 10x + y^2 - 4y = -25$ 20 _____

21 In the diagram below of △PQR, $\overline{ST}$ is drawn parallel to $\overline{PR}$, $PS = 2$, $SQ = 5$, and $TR = 5$.

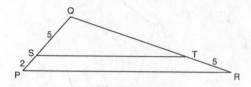

What is the length of $\overline{QR}$?

(1) 7   (3) $12\frac{1}{2}$

(2) 2   (4) $17\frac{1}{2}$ 21 _____

22 The diagram below shows circle O with radii $\overline{OA}$ and $\overline{OB}$. The measure of angle AOB is 120°, and the length of a radius is 6 inches.

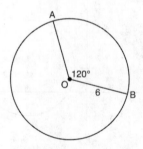

Which expression represents the length of arc AB, in inches?

(1) $\frac{120}{360}(6\pi)$ (3) $\frac{1}{3}(36\pi)$

(2) $120(6)$ (4) $\frac{1}{3}(12\pi)$ 22 _____

23 Line segment CD is the altitude drawn to hypotenuse $\overline{EF}$ in right triangle ECF. If $EC = 10$ and $EF = 24$, then, to the *nearest tenth*, the length of ED is

(1) 4.2   (3) 15.5

(2) 5.4   (4) 21.8   23 _____

24 Line MN is dilated by a scale factor of 2 centered at the point $(0, 6)$. If $\overleftrightarrow{MN}$ is represented by $y = -3x + 6$, which equation can represent $\overleftrightarrow{M'N'}$, the image of $\overleftrightarrow{MN}$?

(1) $y = -3x + 12$ (3) $y = -6x + 12$

(2) $y = -3x + 6$ (4) $y = -6x + 6$ 24 _____

PART II

Answer all 7 questions in this part. Each correct answer will receive 2 credits. Clearly indicate the necessary steps, including appropriate formula substitutions, diagrams, graphs, charts, etc. For all questions in this part, a correct numerical answer with no work shown will receive only 1 credit.   [14 credits]

25 Triangle $A'B'C'$ is the image of triangle ABC after a translation of 2 units to the right and 3 units up. Is triangle ABC congruent to triangle $A'B'C'$? Explain why.

26 Triangle ABC and point $D(1, 2)$ are graphed on the set of axes below.

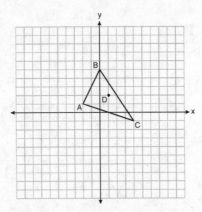

Graph and label $\triangle A'B'C'$, the image of $\triangle ABC$, after a dilation of scale factor 2 centered at point D.

27 Quadrilaterals *BIKE* and *GOLF* are graphed on the set of axes below.

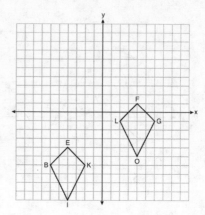

Describe a sequence of transformations that maps quadrilateral *BIKE* onto quadrilateral *GOLF*.

28 In the diagram below, secants $\overline{RST}$ and $\overline{RQP}$, drawn from point R, intersect circle O at S, T, Q, and P.

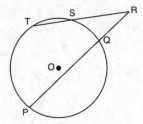

If $RS = 6$, $ST = 4$, and $RP = 15$, what is the length of $\overline{RQ}$?

29 Using a compass and straightedge, construct the median to side $\overline{AC}$ in $\triangle ABC$ below. [Leave all construction marks.]

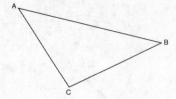

30 Skye says that the two triangles below are congruent. Margaret says that the two triangles are similar.

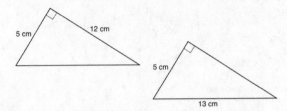

Are Skye and Margaret both correct? Explain why.

31 Randy's basketball is in the shape of a sphere with a maximum circumference of 29.5 inches. Determine and state the volume of the basketball, to the *nearest cubic inch*.

PART III

Answer all 3 questions in this part. Each correct answer will receive 4 credits. Clearly indicate the necessary steps, including appropriate formula substitutions, diagrams, graphs, charts, etc. For all questions in this part, a correct numerical answer with no work shown will receive only 1 credit.   [12 credits]

32 Triangle ABC has vertices with coordinates $A(-1, -1)$, $B(4, 0)$, and $C(0, 4)$. Prove that $\triangle ABC$ is an isosceles triangle but *not* an equilateral triangle. [The use of the set of axes below is optional.]

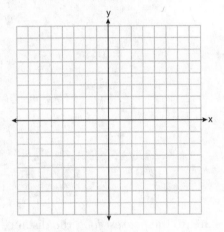

33 The map of a campground is shown below. Campsite C, first aid station F, and supply station S lie along a straight path. The path from the supply station to the tower, T, is perpendicular to the path from the supply station to the campsite. The length of path $\overline{FS}$ is 400 feet. The angle formed by path $\overline{TF}$ and path $\overline{FS}$ is 72°. The angle formed by path $\overline{TC}$ and path $\overline{CS}$ is 55°.

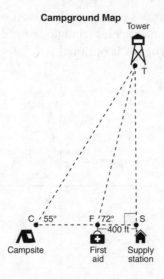

Campground Map

Determine and state, to the *nearest foot*, the distance from the campsite to the tower.

34 Shae has recently begun kickboxing and purchased training equipment as modeled in the diagram below. The total weight of the bag, pole, and unfilled base is 270 pounds. The cylindrical base is 18 inches tall with a diameter of 20 inches. The dry sand used to fill the base weighs 95.46 lbs per cubic foot.

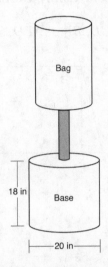

To the *nearest pound*, determine and state the total weight of the training equipment if the base is filled to 85% of its capacity.

PART IV

Answer the question in this part. A correct answer will receive 6 credits. Clearly indicate the necessary steps, including appropriate formula substitutions, diagrams, graphs, charts, etc. A correct numerical answer with no work shown will receive only 1 credit.   **[6 credits]**

35 Given: Parallelogram $ABCD$, $\overline{BF} \perp \overline{AFD}$, and $\overline{DE} \perp \overline{BEC}$

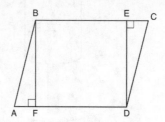

Prove: $BEDF$ is a rectangle

Statements	Reasons
1. Parallelogram $ABCD$, $\overline{BF} \perp \overline{AFD}$, $\overline{DE} \perp \overline{BEC}$	1. Given
2. $\overline{BEC} /\!/ \overline{AFD}$	2. Opposite sides of a parallelogram are parallel
3. $\overline{BE} /\!/ \overline{FD}$	3. Segments contained in parallel lines are parallel
4. $\overline{DE} \perp \overline{AFD}$	4. A segment perpendicular to one of two parallel segments is also perpendicular to the other
5. $\overline{BF} /\!/ \overline{DE}$	5. Two segments perpendicular to the same segment are parallel
6. $BEDF$ is a parallelogram	6. A quadrilateral with opposite sides parallel is a parallelogram
7. $\angle BFD$ is a right angle	7. Perpendicular segments intersect at right angles
8. $BEDF$ is a rectangle	8. A parallelogram with a right angle is a rectangle

Topic	Question Numbers	Number of Points	Your Points	Your Percentage
1. Basic Angle and Segment Relationships	2	2		
2. Angle and Segment Relationships in Triangles and Polygons	18	2		
3. Constructions	29	2		
4. Transformations	1, 3, 19, 25, 26, 27	$2 + 2 + 2 + 2 + 2 + 2 = 12$		
5. Triangle Congruence	30	2		
6. Lines, Segments and Circles on the Coordinate Plane	12, 14, 15, 20, 24	$2 + 2 + 2 + 2 + 2 = 10$		
7. Similarity	4, 5, 9, 11, 21, 23	$2 + 2 + 2 + 2 + 2 + 2 = 12$		
8. Trigonometry	6, 8, 33	$2 + 2 + 4 = 8$		
9. Parallelograms	13, 35	$2 + 6 = 8$		
10. Coordinate Geometry Proofs	32	4		
11. Volume and Solids	7, 10, 16, 31	$2 + 2 + 2 + 2 = 8$		
12. Modeling	34	4		
13. Circles	17, 22, 28	$2 + 2 + 2 = 6$		

HOW TO CONVERT YOUR RAW SCORE TO YOUR GEOMETRY REGENTS EXAMINATION SCORE

The conversion chart below can be used to determine your final score on the June 2018 Regents Exam in Geometry. To find your final exam score, locate in the "Raw Score" column the total number of points you scored out of a possible 86. Then locate in the adjacent column to the right the scale score that corresponds to your raw score. The scale score is your final score.

Raw Score	Scale Score	Performance Level	Raw Score	Scale Score	Performance Level	Raw Score	Scale Score	Performance Level
80	100	5	53	80	4	26	61	2
79	99	5	52	80	4	25	60	2
78	98	5	51	79	3	24	59	2
77	97	5	50	79	3	23	58	2
76	96	5	49	78	3	22	56	2
75	95	5	48	78	3	21	55	2
74	94	5	47	77	3	20	53	1
73	94	5	46	77	3	19	51	1
72	93	5	45	76	3	18	50	1
71	92	5	44	76	3	17	48	1
70	91	5	43	75	3	16	46	1
69	90	5	42	74	3	15	44	1
68	90	5	41	74	3	14	42	1
67	89	5	40	73	3	13	40	1
66	88	5	39	73	3	12	38	1
65	88	5	38	72	3	11	35	1
64	87	5	37	71	3	10	33	1
63	86	5	36	71	3	9	30	1
62	86	5	35	70	3	8	28	1
61	85	5	34	69	3	7	25	1
60	84	4	33	68	3	6	22	1
59	84	4	32	67	3	5	19	1
58	83	4	31	67	3	4	15	1
57	82	4	30	66	3	3	12	1
56	82	4	29	65	3	2	8	1
55	81	4	28	64	2	1	4	1
54	81	4	27	63	2	0	0	1

Answers
June 2018
Geometry

Answer Key

PART I

1. 1	**5.** 4	**9.** 4	**13.** 4	**17.** 3	**21.** 4
2. 3	**6.** 2	**10.** 1	**14.** 2	**18.** 1	**22.** 4
3. 4	**7.** 1	**11.** 2	**15.** 1	**19.** 3	**23.** 1
4. 3	**8.** 1	**12.** 2	**16.** 3	**20.** 2	**24.** 2

PART II

25. $\overline{AB} \cong \overline{A'B'}$, $\overline{BC} \cong \overline{B'C'}$, and so $\overline{CD} \cong \overline{C'D'}$ the triangles are congruent by SSS.

26. $A'(-5, 0)$ $B'(-1, 8)$ $C'(7, -4)$

27. A reflection over the y-axis followed by a translation 5 units up

28. $RQ = 4$

29. Construct the midpoint of side $\overline{AC}$. Connect this midpoint to point B to form the median.

PART III

30. Use the Pythagorean theorem to show the triangles both have side lengths of 5, 12, and 13. The triangles are congruent by SSS and

similar with a scale factor of 1. Skye and Meghan are both correct.

31. Volume $= 434$ in^3

32. Use the distance formula to show $AB = \sqrt{26}$, $BC = \sqrt{32}$, and $CA = \sqrt{26}$. Only two sides are congruent so the triangle is isosceles but not equilateral.

33. The distance from the campsite to the tower is 1503 ft.

34. The total weight is 536 lbs.

PART IV

35. Prove *BEDF* is a parallelogram with a right angle. See the complete solutions for the proof.

In **PARTS II–IV**, you are required to show how you arrived at your answers. For sample methods of solutions, see the *Answers Explained* section.

Answers and Explanations

PART I

1. A rotation is a rigid motion that results in an image congruent to the preimage. All corresponding sides and angles are congruent. By using the given congruence statement $\triangle ABC \cong \triangle RST$, we see that $\angle A$ and $\angle R$ are corresponding congruent angles.

   The correct choice is (**1**).

2. Use the given information to start finding any angle measures by using basic angle relationships. $\angle B$ and $\angle D$ are congruent alternate interior angles formed by the parallel segments, so $m\angle D = 43°$. Since $\angle DEC$ and $\angle FEC$ are a linear pair, they sum to 180°, making $m\angle DEC = 28°$. By applying the angle sum theorem in $\triangle CDE$, we can calculate $m\angle DCE$:

$$m\angle D + m\angle DEC + m\angle DCE = 180°$$
$$43° + 28° + m\angle DCE = 180°$$
$$71° + m\angle DCE = 180°$$
$$m\angle DCE = 109°$$

   $\angle DCE$ and $\angle ACD$ are another linear pair. So they sum to 180°. Calculate $m\angle ACD$:

$$m\angle DCE + m\angle ACD = 180°$$
$$109° + m\angle ACD = 180°$$
$$m\angle ACD = 71°$$

   The correct choice is (**3**).

3. Two reflections over parallel lines are always equivalent to a single translation.

   The correct choice is (**4**).

4. To solve this problem, first show that the two triangles are similar. Then apply the theorem that corresponding sides of similar triangles are proportional. We know $\angle DCA$ and $\angle FCB$ are congruent vertical angles. Next find $m\angle CBF$ using the linear pair relationship:

$$m\angle EBC + m\angle CBF = 180°$$
$$115° + m\angle CBF = 180°$$
$$m\angle CBF = 65°$$

We now know that the two triangles have two pairs of congruent angles. So the triangles are similar by AA. Write a similarity statement by arranging corresponding angles in the same order. A valid similarity statement is $\triangle ADC \sim \triangle FBC$. Now set up a proportion using corresponding sides and cross-multiply:

$$\frac{CB}{CD} = \frac{CF}{AC}$$
$$\frac{CB}{7.2} = \frac{21.6}{9.6}$$
$$(9.6)(CB) = (21.6)(7.2)$$
$$CB = 16.2$$

The correct choice is (**3**).

5. Side lengths are multiplied by the scale factor in a dilation. The dilated square will have side lengths of $9 \times 3 = 27$. The perimeter is equal to the sum of the side lengths. Since a square has four sides, the perimeter is $27 \times 4 = 108$.

The correct choice is (**4**).

6. Sketch the triangle based on the information provided in the question:

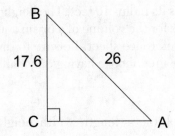

Solve for m∠B using an inverse trig ratio. Relative to ∠B, $\overline{BC}$ is the adjacent side and $\overline{AB}$ is the hypotenuse. Adjacent and hypotenuse indicate the cosine ratio, so use the inverse cosine function:

$$\cos B = \frac{\text{adjacent}}{\text{hypotenuse}}$$

$$\cos B = \frac{17.6}{26}$$

$$B = \cos^{-1}\left(\frac{17.6}{26}\right)$$

$$B = 47°$$

The correct choice is (**2**).

7. The volume of the greenhouse can be found by adding the volume of the prism and the volume of the half-cylinder. The diameter of the half-cylinder is 20 feet, which makes its radius 10 feet. The height of the cylinder is 45 feet. Note that the formulas for the volume of a prism and the volume of a cylinder can be found on the reference sheet. Be sure to divide the cylinder formula by 2 to account for the greenhouse having only half a cylinder.

Rectangular prism:

$$\text{Volume} = \text{length} \cdot \text{width} \cdot \text{height}$$
$$= 45 \text{ ft} \cdot 20 \text{ ft} \cdot 12 \text{ ft}$$
$$= 10{,}800 \text{ ft}^3$$

Half-cylinder:

$$\text{Volume} = \frac{1}{2}\pi r^2 h$$
$$= \frac{1}{2}\pi\left(10^2\right)(45)$$
$$= 7068.58 \text{ ft}^3$$

$$\text{Total Volume} = 10{,}800 \text{ ft}^3 + 7068.58 \text{ ft}^3$$
$$= 17{,}868.58 \text{ ft}^3$$

To the *nearest cubic foot*, the volume is $17{,}869 \text{ ft}^3$.

The correct choice is (**1**).

8. The trigonometric co-function relationship states that if the sine of an angle is equal to the cosine of another angle, the two angles sum to 90°. In this case, the two angle measures are $(2x + 4)$ and 46, so we set their sum equal to 90:

$$2x + 4 + 46 = 90$$
$$2x + 50 = 90$$
$$2x = 40$$
$$x = 20$$

The correct choice is (**1**).

9. When a line is drawn in a triangle parallel to one of the sides, it intercepts a triangle that is similar to the original triangle. In the figure, we see $\triangle ABE$ has $\overline{DFC}$ drawn parallel to side $\overline{AB}$. $\overline{DFC}$ intercepts $\triangle DEF$, making $\triangle DEF$ similar to original triangle $\triangle AEB$.

The correct choice is (**4**).

10. Apply the formula for the volume of a pyramid found on the reference sheet. The area of the rectangular base, represented by B in the formula, is found using area = length $\cdot$ width:

$$\text{Volume} = \frac{1}{3}Bh$$

$$\text{Volume} = \frac{1}{3}(\text{length})(\text{width})h$$

$$82.8 = \frac{1}{3}(9)(4.6)h$$

$$82.8 = (13.8)h$$

$$h = 6$$

The correct choice is (**1**).

11. A line parallel to a side of a triangle will always intercept a smaller triangle that is similar to the original. The two similar triangles are $\triangle ABC$ and $\triangle ADE$. Pairs of corresponding sides are proportional. So the similarity statement $\triangle ABC \sim \triangle ADE$ can be used to write the proportion $\dfrac{AB}{AD} = \dfrac{BC}{DE}$. $\overline{AB}$ and $\overline{AD}$ are formed from the first two vertices of each triangle, while $\overline{BC}$ and $\overline{DE}$ are formed from the second and third vertices of each triangle.

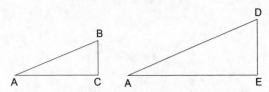

Note that if none of the choices matched using the similarity statement, you could also make valid proportions using the side splitter theorem.

For example, $\dfrac{AB}{BD} = \dfrac{AC}{CE}$ would be a valid proportion.

The correct choice is (**2**).

12. Perpendicular lines have negative reciprocal slopes. The slope of the given line is $\frac{3}{2}$, so the slope of the perpendicular line is $-\frac{2}{3}$. All the choices are in point-slope form, so use that to set up the equation. From the given point $(6, 8)$, use $x_1 = 6$ and $y_1 = 8$:

$$y - y_1 = m(x - x_1)$$
$$y - 8 = -\frac{2}{3}(x - 6)$$

The correct choice is (**2**).

13. A parallelogram can be proven to be a rhombus if it has one of the special rhombus properties:

- Diagonals are perpendicular.
- Consecutive sides are congruent.
- Diagonals bisect the vertex angles.

Choice (4) states that diagonals $\overline{AC}$ and $\overline{BD}$ are perpendicular to each other. None of the other choices would let us conclude a special property from the list above.

The correct choice is (**4**).

14. Find the x- and y-coordinates of F with the following formulas:

$$\text{ratio} = \frac{x - x_1}{x_2 - x} \text{ and ratio} = \frac{y - y_1}{y_2 - y}$$

Using the coordinates $D(-4, -2)$ and $E(1, 8)$, we have $x_1 = -4$, $y_1 = -2$, $x_2 = 1$, and $y_2 = 8$. The ratio for $DF : FE$ is $2 : 3$ or $\frac{2}{3}$.

Solve for the x-coordinate of point F:

$$\text{ratio} = \frac{x - x_1}{x_2 - x}$$

$$\frac{2}{3} = \frac{x - (-4)}{1 - x}$$

$$3(x + 4) = 2(1 - x) \quad \text{Cross-multiply}$$

$$3x + 12 = 2 - 2x$$

$$5x + 12 = 2$$

$$5x = -10$$

$$x = -2$$

Solve for the y-coordinate of point F:

$$\text{ratio} = \frac{y - y_1}{y_2 - y}$$

$$\frac{2}{3} = \frac{y - (-2)}{8 - y}$$

$$3(y + 2) = 2(8 - y) \quad \text{Cross-multiply}$$

$$3y + 6 = 16 - 2y$$

$$5y + 6 = 16$$

$$5y = 10$$

$$y = 2$$

The coordinates of point F are $(-2, 2)$.

The correct choice is (**2**).

15. The easiest method to find the area of △*DAN* is to find the area of a bounding rectangle and then subtract those areas that are not part of △*DAN*. Start by sketching a rectangle that bounds △*DAN* at each vertex. Then divide the bounding rectangle into four regions—three right triangles and △*DAN*.

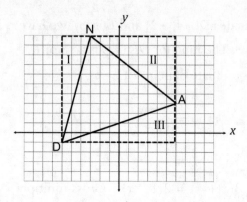

Triangles I, II, and III are each right triangles whose areas can be found using $A = \frac{1}{2}bh$. The bases and heights of triangles I, II, and III are all horizontal and vertical segments whose lengths can be found by counting units along the x- and y-axes. The area of △*DAN* is equal to the area of the bounding rectangle minus the areas of triangles I, II, and III.

Bounding Rectangle:

$$l = 12, \quad w = 11$$
$$A = lw$$
$$= (12)(11)$$
$$= 132$$

Triangle I:

$$b = 3, \quad h = 11$$
$$A = \frac{1}{2}bh$$
$$= \frac{1}{2}(3)(11)$$
$$= 16.5$$

Triangle II:

$$b = 9, \quad h = 7$$

$$A = \frac{1}{2}bh$$

$$= \frac{1}{2}(9)(7)$$

$$= 31.5$$

Triangle III:

$$b = 12, \quad h = 4$$

$$A = \frac{1}{2}bh$$

$$= \frac{1}{2}(12)(4)$$

$$= 24$$

Area of $\triangle DAN$ = Area of rectangle $-$ (Area of $\triangle$I + Area of $\triangle$II + Area of $\triangle$III)

$$= 132 - (16.5 + 31.5 + 24)$$

$$= 60$$

The correct choice is (**1**).

16. A right triangle rotated along one of its legs will always sweep out into a cone. The leg it is rotated around, in this case $\overline{BC}$, is the height of the cone. The other leg, $\overline{AC}$, is the radius. Count units vertically to find $\overline{AC} = 5$, and count units horizontally to find $\overline{BC} = 3$. The height of the cone is 3, and the radius is 5. The diameter is twice the radius, so the diameter is 10. Choice (3) shows a cone with a height of 3 and a diameter of 10.

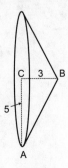

The correct choice is (**3**).

17. The angle formed by two intersecting chords in a circle is always equal to half the sum of the intercepted arcs. The two intercepted arcs are $\overset{\frown}{AC}$ and $\overset{\frown}{BD}$, and the angle formed by the chords is $\angle AEC$:

$$m\angle AEC = \frac{1}{2}\left(m\overset{\frown}{AC} + m\overset{\frown}{BD}\right)$$

$$58° = \frac{1}{2}\left(72° + m\overset{\frown}{BD}\right)$$

$$116° = 72° + m\overset{\frown}{BD}$$

$$m\overset{\frown}{BD} = 44°$$

The correct choice is (**3**).

18. Medians of a triangle intersect at the centroid, which is point M. The centroid divides each median in a 1:2 ratio. The larger part of the median is 2 times the length of the shorter part, and the entire median is 3 times the length of the shorter part. $\overline{MC}$ is the shorter part of median $\overline{SC}$, therefore $SC = 3(MC)$.

The correct choice is (**1**).

19. The minimum number of degrees required to rotate a regular polygon onto itself is given by $\frac{360°}{n}$, where n is the number of sides. The polygon shown has 5 sides, so the minimum rotation angle is $\frac{360°}{5}$, or 72°. Any multiple of 72° will also rotate the polygon onto itself: 72°, 144°, 216°, 288°, and 360°. The only choice that is a multiple of 72° is 216°.

The correct choice is (**3**).

20. Circle O has its center located at $(5, 2)$. Its radius can be found by counting units horizontally or vertically from point O to the circle. The radius is 4 units. Use the center-radius form of the equation of a circle:

$$(x - h)^2 + (y - k)^2 = r^2$$

In this equation, h and k are the coordinates of the center of the circle and r is the radius. Substitute values to write the equation in center-radius form:

$$(x - 5)^2 + (y - 2)^2 = 4^2$$

None of the choices are in this form. So rewrite the equation by expanding each of the squared terms:

$$(x - 5)^2 = x^2 - 10x + 25$$
$$(y - 2)^2 = y^2 - 4y + 4$$

Now rewrite the expanded equation:

$$x^2 - 10x + 25 + y^2 - 4y + 4 = 16$$

Finally, rewrite the equation with all constant terms on the right:

$$x^2 - 10x + y^2 - 4y + 29 = 16$$
$$x^2 - 10x + y^2 - 4y = -13$$

The correct choice is (**2**).

21. The side splitter theorem states that a line parallel to a side of a triangle divides the other two sides proportionally. The four parts of the sides are QS, SP, QT, and TR. Start by writing a proportion involving these four parts. Then solve for QT:

$$\frac{QS}{SP} = \frac{QT}{TR}$$

$$\frac{5}{2} = \frac{QT}{5}$$

$$(5)(5) = 2QT$$

$$QT = 12\frac{1}{2}$$

Now sum QT and TR to get QR:

$$QR = QT + QR$$

$$= 12\frac{1}{2} + 5$$

$$= 17\frac{1}{2}$$

The correct choice is (**4**).

22. Apply the formula for arc length. The central angle θ is 120°, and the radius r is 6:

$$\text{arc length} = \left(\frac{\theta}{360}\right)2\pi r$$

$$= \left(\frac{120}{360}\right)(2\pi)(6)$$

$$= \frac{1}{3}(12\pi)$$

The correct choice is (**4**).

23. An altitude to the hypotenuse of a right triangle will divide the triangle into two similar triangles, which are also similar to the original. We can use that theorem to set up a proportion from the corresponding sides of two of the triangles. Sketch the figure and label the known lengths.

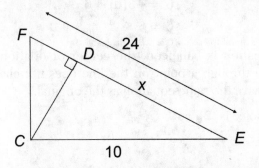

Look at $\triangle CDE$ in the figure. We know the length of hypotenuse $\overline{CE}$ and are looking for the length of longer leg $\overline{ED}$. In $\triangle FCE$, we know the length of both the hypotenuse $\overline{FE}$ and longer leg $\overline{CE}$. Use $\triangle CDE$ and $\triangle FCE$ to set up a proportion and solve for $\overline{ED}$:

$$\frac{\text{hypotenuse}}{\text{long leg}} = \frac{\text{hypotenuse}}{\text{long leg}}$$

$$\frac{CE}{ED} = \frac{FE}{CE}$$

$$\frac{10}{ED} = \frac{24}{10}$$

$$24ED = 100$$

$$ED = 4.1666$$

Rounded to the *nearest tenth*, ED is 4.2.

The correct choice is (**1**).

24. The point $(0, 6)$ lies on line MN. We can confirm this by substituting $x = 0$ and $y = 6$ into the equation of the line:

$$y = -3x + 6$$
$$6 = -3(0) + 6$$
$$6 = 6$$

Since the equation is balanced, the center of dilation lies on the line. Dilating a line through a point on the line does not change the line. The dilation must have the same equation as the original.

The correct choice is (**2**).

PART II

25. A translation is a rigid motion that preserves lengths, so all pairs of corresponding sides are congruent:

$$\overline{AB} \cong \overline{A'B'} \qquad \overline{BC} \cong \overline{B'C'} \qquad \overline{CD} \cong \overline{C'D'}$$

Since all three pairs of sides are congruent, **the two triangles are congruent by the SSS postulate**.

26. A dilation increases the size of the original figure by the scale factor. It also increases the distance of each point from the center. We can use this second feature of a dilation to find the coordinates of A', B', and C'. To find A', count the number of horizontal and vertical units from D to A. Point A is 3 units left and 1 unit down from point D. Since we need a scale factor of 2, double those numbers to get 6 units left and 2 units down. Counting 6 left and 2 down from D puts A' at $(-5, 0)$. Repeat this process to find B' and C'.

$$D \text{ to } A: 3 \text{ left, 1 down}$$
$$D \text{ to } A': 6 \text{ left, 2 down}$$

$$D \text{ to } B: 1 \text{ left, 3 up}$$
$$D \text{ to } B': 2 \text{ left, 6 up}$$

$$D \text{ to } C: 3 \text{ right, 3 down}$$
$$D \text{ to } C': 6 \text{ right, 6 down}$$

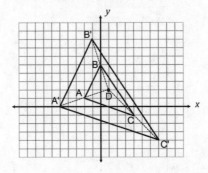

The new coordinates are **$A'(-5, 0)$, $B'(-1, 8)$, $C'(7, -4)$**.

27. A good strategy for any problem that asks you to find a sequence of transformations is first to identify if a reflection is needed. The orientation is the direction, clockwise or counterclockwise, that the vertices are encountered. If the orientation changes, then a reflection is required since only a reflection can change the orientation. *BIKE* has a counterclockwise orientation, while *GOLF* has a clockwise orientation. Therefore, we need a reflection. Each figure is symmetric about a vertical line, so we can reflect about the y-axis. Following the reflection, we see that we just need to translate $B'I'K'E'$ up 5 units to map it onto *GOLF*.

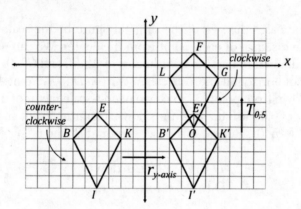

The complete sequence is a **reflection over the y-axis followed by a translation 5 units up, or $T_{0,5} \circ r_{y\text{-axis}}$**.

28. When two secants intersect, the relationship between the parts of the secants can be shown by the following formula:

$$\text{Outside} \cdot \text{Whole} = \text{Outside} \cdot \text{Whole}$$

The outside of each secant is the part that lies outside the circle, in this case $\overline{RS}$ and $\overline{RQ}$. The whole part represents the entire length of the secant, or $\overline{RT}$ and $\overline{RP}$. To find the length of $\overline{RQ}$, apply the formula, using the sum of $RS + ST$ for RT:

$$\text{Outside} \cdot \text{Whole} = \text{Outside} \cdot \text{Whole}$$
$$RS \cdot RT = RQ \cdot RT$$
$$6 \cdot 10 = RQ \cdot 15$$
$$60 = 15RQ$$
$$RQ = 4$$

The length of $\overline{RQ}$ is 4.

29. A median of a triangle is a segment from a vertex to the opposite midpoint. Begin by constructing the midpoint of $\overline{AC}$. With the point of the compass at A, open the compass more than half the distance of $\overline{AC}$ and make an arc. Without changing the compass setting, move the compass point to C. Make another arc that intersects the first arc as shown in figure (a). Connect the two points of intersection using a straightedge. Figure (b) shows this new segment intersecting $\overline{AC}$ at midpoint D. Finally, use the straightedge to draw a segment from point B to midpoint D. This is shown in figure (c). Segment $\overline{BD}$ is the median to side $\overline{AC}$.

(a)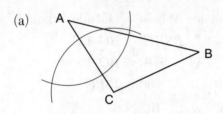

(b)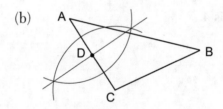

(c)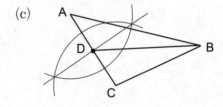

PART III

30. Triangles can be proven to be similar using the AA, SAS, or SSS postulates. In this example, apply the Pythagorean theorem to each triangle to find the missing side. Then check if each pair of corresponding sides are in the same ratio. If they are, then the triangles are similar by SSS.

Left triangle:

$$a^2 + b^2 = c^2$$
$$5^2 + 12^2 = c^2$$
$$25 + 144 = c^2$$
$$169 = c^2$$
$$c = 13$$

Right triangle:

$$a^2 + b^2 = c^2$$
$$5^2 + b^2 = 13^2$$
$$25 + b^2 = 169$$
$$b^2 = 144$$
$$b = 12$$

The three sides of the left triangle are 5, 12, and 13. The three sides of the right triangle are also 5, 12, and 13. All three pairs of sides are in the same ratio:

$$\frac{5}{5} = \frac{12}{12} = \frac{13}{13} = 1$$

Since all three pairs of sides are in the same ratio, **the triangles are similar by SSS. The triangles are also congruent by the SSS postulate because all three pairs of sides are congruent. Therefore, both Skye and Margaret are correct.**

31. The radius of the basketball is needed to find the volume. Use the given circumference to calculate the radius. The circumference formula can be found on the reference sheet:

$$C = 2\pi r$$
$$29.5 = 2\pi r$$
$$r = \frac{29.5}{2\pi}$$
$$= 4.695070821 \text{ in.}$$

Now apply the formula for the volume of a sphere, which is also found on the reference sheet:

$$V = \frac{4}{3}\pi r^3$$
$$V = \frac{4}{3}\pi (4.695070821)^3$$
$$= 433.5259036 \text{ in.}^3$$

The volume of the basketball to the *nearest cubic inch* is **434 in.**^{3}

32. Use the distance formula to find the length of each side of the triangle. If two sides have the same length and the third length is different, we can conclude the triangle is isosceles but not equilateral:

$$d = \sqrt{\left(x_2 - x_1\right)^2 + \left(y_2 - y_1\right)^2}$$

$$AB = \sqrt{\left(4 - (-1)\right)^2 + \left(0 - (-1)\right)^2}$$

$$= \sqrt{5^2 + 1^2}$$

$$= \sqrt{26}$$

$$BC = \sqrt{\left(0 - 4\right)^2 + \left(4 - 0\right)^2}$$

$$= \sqrt{\left(-4\right)^2 + 4^2}$$

$$= \sqrt{32}$$

$$CA = \sqrt{\left(-1 - 0\right)^2 + \left(-1 - 4\right)^2}$$

$$= \sqrt{\left(-1\right)^2 + \left(-5\right)^2}$$

$$= \sqrt{26}$$

$\triangle ABC$ is isosceles because $AB = CA$. $\triangle ABC$ is not equilateral because BC is not equal to the other two side lengths.

33. Start by sketching the figure as two separate right triangles.

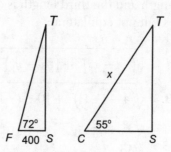

The distance from the campsite to the tower is side length CT in $\triangle CST$. We can apply a trig ratio to find the length of ST in $\triangle FST$. Then the length of CT can be found using a trig ratio in $\triangle CST$.

Find ST using $\triangle FST$. Relative to $\angle F$, $\overline{ST}$ is the opposite and $\overline{FS}$ is the adjacent. Therefore, use the tangent ratio:

$$\text{tangent} = \frac{\text{opposite}}{\text{adjacent}}$$

$$\tan F = \frac{ST}{FS}$$

$$\tan 72° = \frac{ST}{400}$$

$$ST = (400)(\tan 72°) \quad \text{Cross-multiply}$$

$$ST = 1231.073415$$

Find CT using $\triangle CST$. Relative to $\angle C$, $\overline{ST}$ is the opposite and $\overline{CT}$ is the hypotenuse. Therefore, use the sine ratio:

$$\text{sine} = \frac{\text{opposite}}{\text{hypotenuse}}$$

$$\sin C = \frac{ST}{CT}$$

$$\sin 55° = \frac{1231.073415}{CT}$$

$$(CT)(\sin 55°) = 1231.073415 \quad \text{Cross-multiply}$$

$$CT = \frac{1231.073415}{\sin 55°}$$

$$CT = 1502.863142$$

Rounded to the *nearest foot*, the distance from the campsite to the tower is **1503 feet**.

34. The total weight of the equipment is the given weight of the bag, pole, and unfilled base plus the weight of the sand in the base. Find the weight of the sand by first calculating the volume of the base and by then applying the following relationship:

$$\text{weight} = \text{density} \cdot \text{volume}$$

The given density is in lbs per cubic foot, while the dimensions of the base are in inches. Convert the base's dimensions to feet in order to have consistent units. Remember that 1 foot is equal to 12 inches:

$$\text{height} = 18 \text{ in.} \cdot \frac{1 \text{ ft}}{12 \text{ in.}}$$

$$= 1.5 \text{ ft}$$

$$\text{diameter} = 20 \text{ in.} \cdot \frac{1 \text{ ft}}{12 \text{ in.}}$$

$$= 1.666666667 \text{ ft}$$

Find the volume of the base using the volume formula of a cylinder found on the reference sheet. Remember to divide the diameter by 2 to get the radius:

$$V = \pi r^2 h$$
$$= \pi (0.833333333 \text{ ft})^2 (1.5 \text{ ft})$$
$$= 3.272492347 \text{ ft}^3$$

Calculate the mass of the sand, multiplying by 0.85 to account for the base not being filled completely:

$$\text{weight} = \text{density} \cdot \text{volume} \cdot 0.85$$
$$= 95.46 \frac{\text{lbs}}{\text{ft}^3} \cdot 3.272492347 \text{ ft}^3 \cdot 0.85$$
$$= 265.5333015 \text{ lbs}$$

Sum the weight of the sand and the equipment:

$$\text{total weight} = 265.5333015 \text{ lbs} + 270 \text{ lbs}$$
$$= 535.5333015$$

Rounded to the *nearest pound*, the total weight of the training equipment is **536 lbs**.

PART IV

35. A parallelogram can be proven to be a rectangle if it has one of the special properties of rectangles. These properties are right angles (just one right angle need be proven) and congruent diagonals. Since we are given parallel and perpendicular segments, it makes sense to prove *BEDF* is a parallelogram by showing it has opposite sides parallel and then show that *BEDF* has a right angle. The key to showing that opposite sides are parallel is to realize that $\overline{BF}$ and $\overline{ED}$ are both perpendicular to $\overline{FD}$, so they must be parallel to each other.

Examination
August 2018
Geometry

HIGH SCHOOL MATH REFERENCE SHEET

Conversions

1 inch = 2.54 centimeters

1 meter = 39.37 inches

1 mile = 5280 feet

1 mile = 1760 yards

1 mile = 1.609 kilometers

1 kilometer = 0.62 mile

1 pound = 16 ounces

1 pound = 0.454 kilogram

1 kilogram = 2.2 pounds

1 ton = 2000 pounds

1 cup = 8 fluid ounces

1 pint = 2 cups

1 quart = 2 pints

1 gallon = 4 quarts

1 gallon = 3.785 liters

1 liter = 0.264 gallon

1 liter = 1000 cubic centimeters

Formulas

Triangle	$A = \frac{1}{2}bh$
Parallelogram	$A = bh$
Circle	$A = \pi r^2$
Circle	$C = \pi d$ or $C = 2\pi r$

Formulas (continued)

General Prisms $\qquad V = Bh$

Cylinder $\qquad V = \pi r^2 h$

Sphere $\qquad V = \dfrac{4}{3}\pi r^3$

Cone $\qquad V = \dfrac{1}{3}\pi r^2 h$

Pyramid $\qquad V = \dfrac{1}{3}Bh$

Pythagorean Theorem $\qquad a^2 + b^2 = c^2$

Quadratic Formula $\qquad x = \dfrac{-b \pm \sqrt{b^2 - 4ac}}{2a}$

Arithmetic Sequence $\qquad a_n = a_1 + (n-1)d$

Geometric Sequence $\qquad a_n = a_1 r^{n-1}$

Geometric Series $\qquad S_n = \dfrac{a_1 - a_1 r^n}{1-r}$ where $r \neq 1$

Radians $\qquad 1 \text{ radian} = \dfrac{180}{\pi}\text{ degrees}$

Degrees $\qquad 1 \text{ degree} = \dfrac{\pi}{180}\text{ radians}$

Exponential Growth/Decay $\qquad A = A_0 e^{k(t-t_0)+B_0}$

180 Minutes–35 Questions

PART I

Answer all 24 questions in this part. Each correct answer will receive 2 credits. No partial credit will be allowed. For each statement or question, write in the space provided, the numeral preceding the word or expression that best completes the statement or answers the question. [48 credits]

1 In the diagram below, $\overline{AEFB} \parallel \overline{CGD}, \overline{GE},$ and $\overline{GF}$ are drawn.

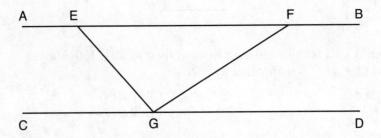

If m∠EFG = 32° and m∠AEG = 137°, what is m∠EGF?

(1) 11°   (3) 75°

(2) 43°   (4) 105°   1 _____

2 If △ABC is mapped onto △DEF after a line reflection and △DEF is mapped onto △XYZ after a translation, the relationship between △ABC and △XYZ is that they are always

(1) congruent and similar

(2) congruent but not similar

(3) similar but not congruent

(4) neither similar nor congruent   2 _____

3 An isosceles right triangle whose legs measure 6 is continuously rotated about one of its legs to form a three-dimensional object. The three-dimensional object is a

(1) cylinder with a diameter of 6

(2) cylinder with a diameter of 12

(3) cone with a diameter of 6

(4) cone with a diameter of 12   3 _____

4 In regular hexagon *ABCDEF* shown below, $\overline{AD}$, $\overline{BE}$, and $\overline{CF}$ all intersect at *G*.

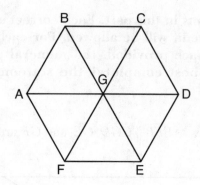

When △*ABG* is reflected over $\overline{BG}$ and then rotated 180° about point *G*, △*ABG* is mapped onto

(1) △*FEG*   (3) △*CBG*

(2) △*AFG*   (4) △*DEG*   4____

5 A right cylinder is cut perpendicular to its base. The shape of the cross section is a

(1) circle   (3) rectangle

(2) cylinder   (4) triangular prism   5____

6 Yolanda is making a springboard to use for gymnastics. She has 8-inch-tall springs and wants to form a 16.5° angle with the base, as modeled in the diagram below.

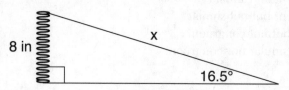

To the *nearest tenth of an inch*, what will be the length of the springboard, *x*?

(1) 2.3   (3) 27.0

(2) 8.3   (4) 28.2   6____

7 In the diagram below of right triangle ABC, altitude $\overline{BD}$ is drawn
to hypotenuse $\overline{AC}$.

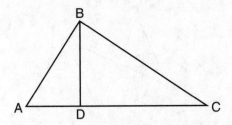

If $BD = 4$, $AD = x - 6$, and $CD = x$, what is the length of $\overline{CD}$?

(1) 5   (3) 8
(2) 2   (4) 11   7____

8 Rhombus $STAR$ has vertices $S(-1, 2)$, $T(2, 3)$, $A(3, 0)$, and $R(0, -1)$.
What is the perimeter of rhombus $STAR$?

(1) $\sqrt{34}$ (3) $\sqrt{10}$
(2) $4\sqrt{34}$ (4) $4\sqrt{10}$ 8____

9 In the diagram below of $\triangle HAR$ and $\triangle NTY$, angles H and N are
right angles, and $\triangle HAR \sim \triangle NTY$.

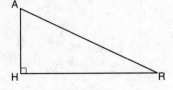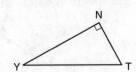

If $AR = 13$ and $HR = 12$, what is the measure of angle Y, to the
nearest degree?

(1) 23°   (3) 65°
(2) 25°   (4) 67°   9____

10 In the diagram below, $\overline{AKS}$, $\overline{NKC}$, $\overline{AN}$, and $\overline{SC}$ are drawn such that $AN \cong SC$.

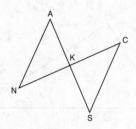

Which additional statement is sufficient to prove $\triangle KAN \cong \triangle KSC$ by AAS?

(1) $\overline{AS}$ and $\overline{NC}$ bisect each other.   (3) $\overline{AS} \perp \overline{CN}$

(2) K is the midpoint of $\overline{NC}$.   (4) $\overline{AN} \parallel \overline{SC}$ 10 _____

11 Which equation represents a line that is perpendicular to the line represented by $y = \frac{2}{3}x + 1$?

(1) $3x + 2y = 12$ (3) $y = \frac{3}{2}x + 2$

(2) $3x - 2y = 12$ (4) $y = -\frac{2}{3}x + 4$ 11 _____

12 In the diagram of $\triangle ABC$ below, points D and E are on sides $\overline{AB}$ and $\overline{CB}$ respectively, such that $\overline{DE} \parallel \overline{AC}$.

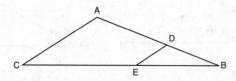

If EB is 3 more than DB, $AB = 14$, and $CB = 21$, what is the length of $\overline{AD}$?

(1) 6   (3) 9

(2) 8   (4) 12   12 _____

13 Quadrilateral *MATH* has both pairs of opposite sides congruent and parallel. Which statement about quadrilateral *MATH* is always true?

  (1) $\overline{MT} \cong \overline{AH}$ (3) $\angle MHT \cong \angle ATH$

  (2) $\overline{MT} \perp \overline{AH}$ (4) $\angle MAT \cong \angle MHT$ 13 _____

14 In the figure shown below, quadrilateral *TAEO* is circumscribed around circle *D*. The midpoint of $\overline{TA}$ is *R*, and $\overline{HO} \cong \overline{PE}$.

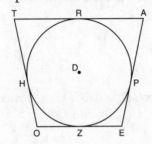

If $AP = 10$ and $EO = 12$, what is the perimeter of quadrilateral *TAEO*?

  (1) 56   (3) 72

  (2) 64   (4) 76   14 _____

15 The coordinates of the endpoints of directed line segment *ABC* are $A(-8, 7)$ and $C(7, -13)$. If $AB:BC = 3:2$, the coordinates of *B* are

  (1) $(1, -5)$ (3) $(-3, 0)$

  (2) $(-2, -1)$ (4) $(3, -6)$ 15 _____

16 In triangle ABC, points D and E are on sides $\overline{AB}$ and $\overline{BC}$, respectively, such that $\overline{DE} \parallel \overline{AC}$, and $AD{:}DB = 3{:}5$.

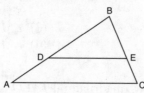

If $DB = 6.3$ and $AC = 9.4$, what is the length of $\overline{DE}$, to the *nearest tenth*?

(1) 3.8   (3) 5.9
(2) 5.6   (4) 15.7   16 _____

17 In the diagram below, rectangle $ABCD$ has vertices whose coordinates are $A(7, 1)$, $B(9, 3)$, $C(3, 9)$, and $D(1, 7)$.

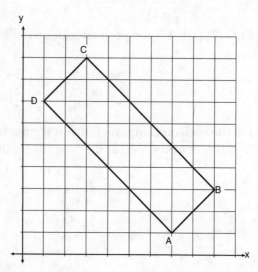

Which transformation will *not* carry the rectangle onto itself?

(1) a reflection over the line $y = x$
(2) a reflection over the line $y = -x + 10$
(3) a rotation of $180°$ about the point $(6, 6)$
(4) a rotation of $180°$ about the point $(5, 5)$ 17 _____

18 A circle with a diameter of 10 cm and a central angle of 30° is drawn below.

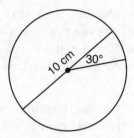

What is the area, to the *nearest tenth of a square centimeter*, of the sector formed by the 30° angle?

(1) 5.2   (3) 13.1

(2) 6.5   (4) 26.2   18 _____

19 A child's tent can be modeled as a pyramid with a square base whose sides measure 60 inches and whose height measures 84 inches. What is the volume of the tent, to the *nearest cubic foot*?

(1) 35   (3) 82

(2) 58   (4) 175   19 _____

20 In the accompanying diagram of right triangle ABC, altitude $\overline{BD}$ is drawn to hypotenuse $\overline{AC}$.

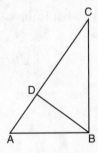

Which statement must always be true?

(1) $\dfrac{AD}{AB} = \dfrac{BC}{AC}$ (3) $\dfrac{BD}{BC} = \dfrac{AB}{AD}$

(2) $\dfrac{AD}{AB} = \dfrac{AB}{AC}$ (4) $\dfrac{AB}{BC} = \dfrac{BD}{AC}$ 20 _____

21 An equation of circle O is $x^2 + y^2 + 4x - 8y = -16$. The statement that best describes circle O is the

(1) center is $(2, -4)$ and is tangent to the x-axis
(2) center is $(2, -4)$ and is tangent to the y-axis
(3) center is $(-2, 4)$ and is tangent to the x-axis
(4) center is $(-2, 4)$ and is tangent to the y-axis   21 _____

22 In $\triangle ABC$, $\overline{BD}$ is the perpendicular bisector of $\overline{ADC}$. Based upon this information, which statements below can be proven?

   I. $\overline{BD}$ is a median.
  II. $\overline{BD}$ bisects $\angle ABC$.
 III. $\triangle ABC$ is isosceles.

(1) I and II, only   (3) II and III, only
(2) I and III, only   (4) I, II, and III   22 _____

23 Triangle RJM has an area of 6 and a perimeter of 12. If the triangle is dilated by a scale factor of 3 centered at the origin, what are the area and perimeter of its image, triangle $R'J'M'$?

(1) area of 9 and perimeter of 15
(2) area of 18 and perimeter of 36
(3) area of 54 and perimeter of 36
(4) area of 54 and perimeter of 108   23 _____

24 If $\sin (2x + 7)° = \cos (4x - 7)°$, what is the value of x?

(1) 7   (3) 21
(2) 15   (4) 30   24 _____

PART II

Answer all 7 questions in this part. Each correct answer will receive 2 credits. Clearly indicate the necessary steps, including appropriate formula substitutions, diagrams, graphs, charts, etc. For all questions in this part, a correct numerical answer with no work shown will receive only 1 credit.   [14 credits]

25 In the circle below, $\overline{AB}$ is a chord. Using a compass and straightedge, construct a diameter of the circle. [Leave all construction marks.]

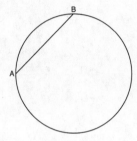

26 In parallelogram $ABCD$ shown below, the bisectors of $\angle ABC$ and $\angle DCB$ meet at E, a point on $\overline{AD}$.

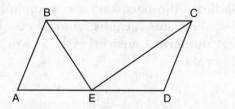

If $m\angle A = 68°$, determine and state $m\angle BEC$.

27 In circle A below, chord $\overline{BC}$ and diameter $\overline{DAE}$ intersect at F.

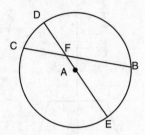

If $m\widehat{CD} = 46°$ and $m\widehat{DB} = 102°$, what is $m\angle CFE$?

28 Trapezoids *ABCD* and *A″ B″ C″ D″* are graphed on the set of axes below.

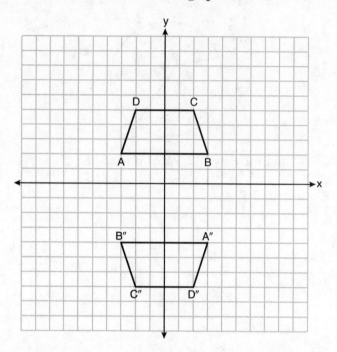

Describe a sequence of transformations that maps trapezoid *ABCD* onto trapezoid *A″ B″ C″ D″*.

29 In the model below, a support wire for a telephone pole is attached to the pole and anchored to a stake in the ground 15 feet from the base of the telephone pole. Jamal places a 6-foot wooden pole under the support wire parallel to the telephone pole, such that one end of the pole is on the ground and the top of the pole is touching the support wire. He measures the distance between the bottom of the pole and the stake in the ground.

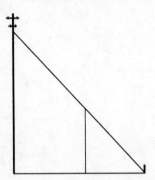

Jamal says he can approximate how high the support wire attaches to the telephone pole by using similar triangles. Explain why the triangles are similar.

30 Aliyah says that when the line $4x + 3y = 24$ is dilated by a scale factor of 2 centered at the point $(3, 4)$, the equation of the dilated line is $y = -\dfrac{4}{3}x + 16$. Is Aliyah correct? Explain why.

[The use of the set of axes below is optional.]

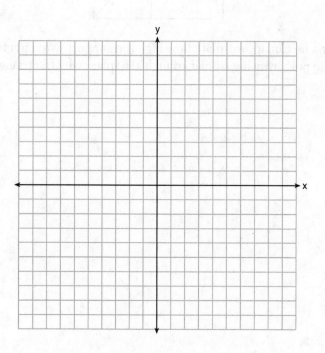

31 Ian needs to replace two concrete sections in his sidewalk, as modeled below. Each section is 36 inches by 36 inches and 4 inches deep. He can mix his own concrete for $3.25 per cubic foot.

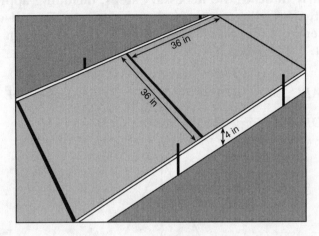

How much money will it cost Ian to replace the two concrete sections?

PART III

Answer all 3 questions in this part. Each correct answer will receive 4 credits. Clearly indicate the necessary steps, including appropriate formula substitutions, diagrams, graphs, charts, etc. For all questions in this part, a correct numerical answer with no work shown will receive only 1 credit. [12 credits]

32 Given: $\triangle ABC$, $\overline{AEC}$, $\overline{BDE}$ with $\angle ABE \cong \angle CBE$, and $\angle ADE \cong \angle CDE$

Prove: $\overline{BDE}$ is the perpendicular bisector of $\overline{AC}$

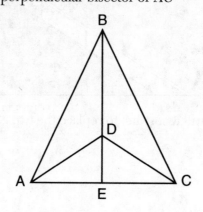

Fill in the missing statement and reasons below.

Statements	Reasons
(1) $\triangle ABC$, $\overline{AEC}$, $\overline{BDE}$ with $\angle ABE \cong \angle CBE$ and $\angle ADE \cong \angle CDE$	(1) Given
(2) $\overline{BD} \cong \overline{BD}$	(2) _____
(3) $\angle BDA$ and $\angle ADE$ are supplementary. $\angle BDC$ and $\angle CDE$ are supplementary.	(3) Linear pairs of angles are supplementary.
(4) _____	(4) Supplements of congruent angles are congruent.
(5) $\triangle ABD \cong \triangle CBD$	(5) ASA
(6) $\overline{AD} \cong \overline{CD}$, $\overline{AB} \cong \overline{CB}$	(6) _____
(7) $\overline{BDE}$ is the perpendicular bisector of $\overline{AC}$.	(7) _____ _____ _____

33 A homeowner is building three steps leading to a deck, as modeled by the diagram below. All three step rises, $\overline{HA}$, $\overline{FG}$, and $\overline{DE}$, are congruent, and all three step runs, $\overline{HG}$, $\overline{FE}$, and $\overline{DC}$, are congruent. Each step rise is perpendicular to the step run it joins. The measure of $\angle CAB = 36°$ and m$\angle CBA = 90°$.

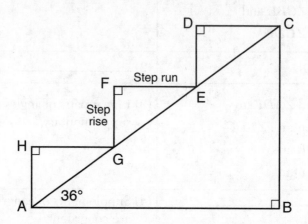

If each step run is parallel to $\overline{AB}$ and has a length of 10 inches, determine and state the length of each step rise, to the *nearest tenth of an inch*.

Determine and state the length of $\overline{AC}$, to the *nearest inch*.

34 A bakery sells hollow chocolate spheres. The larger diameter of each sphere is 4 cm. The thickness of the chocolate of each sphere is 0.5 cm. Determine and state, to the *nearest tenth of a cubic centimeter*, the amount of chocolate in each hollow sphere.

The bakery packages 8 of them into a box. If the density of the chocolate is 1.308 g/cm^3, determine and state, to the *nearest gram*, the total mass of the chocolate in the box.

PART IV

Answer the question in this part. Each correct answer will receive 6 credits. Clearly indicate the necessary steps, including appropriate formula substitutions, diagrams, graphs, charts, etc. For all questions in this part, a correct numerical answer with no work shown will receive only 1 credit. [6 credits]

35 The vertices of quadrilateral *MATH* have coordinates $M(-4, 2)$, $A(-1, -3)$, $T(9, 3)$, and $H(6, 8)$.

Prove that quadrilateral *MATH* is a parallelogram.

[The use of the set of axes on the next page is optional.]

Question 35 is continued on the next page.

Question 35 continued

Prove that quadrilateral *MATH* is a rectangle.

[The use of the set of axes below is optional.]

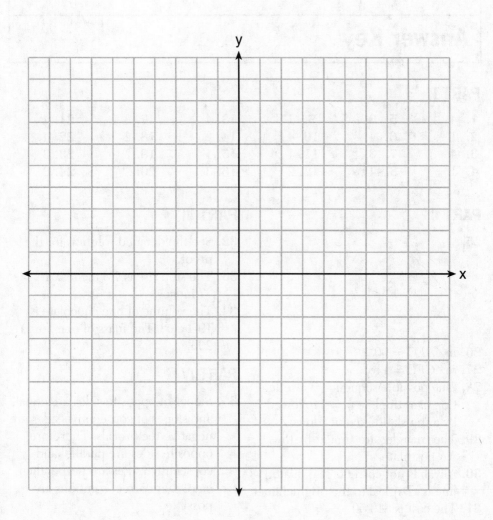

Answers
August 2018
Geometry

Answer Key

PART I

1. 4	**5.** 3	**9.** 1	**13.** 4	**17.** 3	**21.** 4
2. 1	**6.** 4	**10.** 4	**14.** 2	**18.** 2	**22.** 4
3. 4	**7.** 3	**11.** 1	**15.** 1	**19.** 2	**23.** 3
4. 1	**8.** 4	**12.** 2	**16.** 3	**20.** 2	**24.** 2

PART II

25.

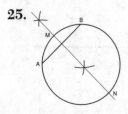

26. m∠BEC = 90°

27. m∠CFE = 118°

28. One possible sequence is a reflection over the line $y = 2$ followed by a translation 6 units down.

29. The triangles are similar by the AA postulate.

30. Aliyah is not correct. The dilated line is identical to the original line.

31. The cost is $19.50

PART III

32. See the detailed solution for the proof.

33. The step rise is 7.3 inches. $AC = 37$ inches.

34. The volume of one chocolate is 19.4 cm^3. The mass of one box is 203 g.

PART IV

35. Slopes of opposite sides are equal, and slopes of consecutive sides are negative reciprocals. Therefore, opposite sides are parallel and consecutive sides are perpendicular. Other correct solutions are possible.

In **PARTS II–IV**, you are required to show how you arrived at your answers. For sample methods of solutions, see the *Answers Explained* section.

Answers Explained

PART I

1. $\angle AEG$ and $\angle GEF$ are a linear pair, which sum to $180°$.

$$m\angle AEG + m\angle GEF = 180°$$
$$137° + m\angle GEF = 180°$$
$$m\angle GEF = 43°$$

Next, apply the angle sum theorem in $\triangle EFG$.

$$m\angle GEF + m\angle EFG + m\angle EGF = 180°$$
$$43° + 32° + m\angle EGF = 180°$$
$$75° + m\angle EGF = 180°$$
$$m\angle EGF = 105°$$

The correct choice is (**4**).

2. Line reflections and translations are both rigid motions that result in congruent images. Therefore, $\triangle ABC \cong \triangle DEF$ and $\triangle DEF \cong \triangle XYZ$, which implies that $\triangle ABC \cong \triangle XYZ$. Congruent images are also always similar with a scale factor of 1. $\triangle ABC$ and $\triangle XYZ$ are both congruent and similar.

The correct choice is (**1**).

3. A right triangle rotated about one of its legs will always form a cone, with the radius of the cone equal to one leg length and the height of the cone equal to the other leg length. In this case, the height of the cone is 6 and the radius is 6, making the diameter 12.

The correct choice is (**4**).

4. The reflection over $\overline{BG}$ maps $\triangle ABG$ onto $\triangle CBG$. The 180° rotation then maps it onto $\triangle FEG$.

The correct choice is (**1**).

5. The cross section of a right cylinder cut perpendicular to its base is always a rectangle. It may also be a square if the diameter is equal to the height.

The correct choice is (**3**).

6. We have a right triangle with a given angle and side and are looking for a second side. Relative to the 16.5° angle, the unknown length x is the hypotenuse and the 8-inch length is the opposite. The sine ratio can be used to solve for x.

$$\sin\left(16.5°\right) = \frac{opposite}{hypotenuse}$$
$$\sin\left(16.5°\right) = \frac{8}{x}$$
$$x = \frac{8}{\sin\left(16.5°\right)}$$
$$x = 28.16749$$

To the nearest tenth of an inch, $x = 28.2$.

The correct choice is (**4**).

7. When an altitude is drawn to the hypotenuse of a right triangle, the altitude is the mean proportional to the parts of the hypotenuse. The relationship in this case is:

$$\frac{AD}{BD} = \frac{BD}{CD}$$

$$\frac{x - 6}{4} = \frac{4}{x}$$

Next, cross multiply and solve the quadratic equation by factoring.

$$x(x - 6) = 4 \times 4$$

$$x^2 - 6x = 16$$

$$x^2 - 6x - 16 = 0$$

$$(x - 8)(x + 2) = 0$$

$$x - 8 = 0 \qquad x + 2 = 0$$

$$x = 8 \qquad x = -2$$

There are two solutions to the quadratic equation, but the length of CD cannot be negative, so the only possible answer is $x = 8$.

The correct choice is (**3**).

8. The perimeter of a polygon is the sum of the lengths of all its sides. Since a rhombus has four congruent sides, we need to find the length of only one side and then multiply that by 4. The length of a side can be found using the distance formula. Using the coordinates for side $\overline{ST}$, we have:

$$ST = \sqrt{(x_2 - x_1)^2 + (y_2 - y_1)^2}$$

$$= \sqrt{(2 - (-1))^2 + (3 - 2)^2}$$

$$= \sqrt{(3)^2 + (1)^2}$$

$$= \sqrt{10}$$

The perimeter is $4\sqrt{10}$.

The correct choice is (**4**).

9. From the similarity statement $\triangle HAR \sim \triangle NTY$, we know that corresponding angles $\angle R$ and $\angle Y$ are congruent. We can find $\angle R$ using an inverse trigonometric ratio. Relative to $\angle R$, $\overline{AR}$ is the hypotenuse and $\overline{HR}$ is the adjacent, so we need an inverse cosine function.

$$\cos(R) = \frac{adjacent}{hypotenuse}$$

$$\cos(R) = \frac{12}{13}$$

$$m\angle R = \cos^{-1}\frac{12}{13}$$

$$m\angle R = 22.619°$$

Rounded to the nearest degree, $m\angle R = 23°$ and $m\angle Y = 23°$.

The correct choice is **(1)**.

10. The given congruent sides $\overline{AN}$ and $\overline{SC}$, along with congruent vertical angles $\angle AKN$ and $\angle CKS$, provide one pair of sides and one pair of angles. To use AAS, we need one more pair of angles. Choices (1) and (2) lead us to congruent sides, not angles, so those can be eliminated. Choice (3) would let us conclude that $\angle AKN$ and $\angle CKS$ are congruent right angles, but that does not provide the additional pair of congruent angles needed since we already know those angles are congruent. Choice (4) tells us that $\overline{AN}$ and $\overline{SC}$ are parallel. The parallel sides let us conclude two pairs of congruent alternate interior angles, $\angle N \cong \angle C$ and $\angle A \cong \angle S$, which gives us sufficient information to use the AAS postulate.

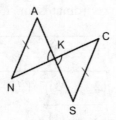

The correct choice is **(4)**.

11. Perpendicular lines have negative reciprocal slopes. The slope of the given line is $\frac{2}{3}$, so the slope of a perpendicular line must be $-\frac{3}{2}$. Rearrange each equation into $y = mx + b$ form to find the slope if necessary.

Choice (1):

$$3x + 2y = 12$$
$$2y = -3x + 12$$
$$y = -\frac{3}{2}x + 6$$

This has the correct slope, so the correct choice is (**1**).

12. First, the statement "*EB* is 3 more than *DB*" needs to be converted into an algebraic expression. If we let $DB = x$, then *EB* is $x + 3$. A line parallel to a side of a triangle forms two similar triangles, so $\triangle ABC \sim \triangle DBE$. We can use the fact that pairs of corresponding sides are proportional to solve for x. Sketching the triangles separately helps identify the parts we want to use in our proportion.

$$\frac{DB}{AB} = \frac{EB}{CB}$$
$$\frac{x}{14} = \frac{x+3}{21}$$
$$21x = 14(x+3)$$
$$21x = 14x + 42$$
$$7x = 42$$
$$x = 6$$

Next find *AD*.

$$AD = AB - DB$$
$$= 14 - 6$$
$$= 8$$

The correct choice is (**2**).

13. A quadrilateral with opposite sides congruent and parallel is a parallelogram. Evaluate each choice to determine which one is consistent with the properties of a parallelogram.

- Choice (1): $\overline{MT} \cong \overline{AH}$ states that the diagonals are congruent. This is true of rectangles but not all parallelograms.
- Choice (2): $\overline{MT} \cong \overline{AH}$ states that the diagonals are perpendicular. This is true of rhombuses but not all parallelograms.
- Choice (3): $\angle MHT \cong \angle ATH$ states that a pair of consecutive angles are congruent. This is true of rectangles but not all parallelograms.
- Choice (4): $\angle MAT \cong \angle MHT$ states that a pair of opposite angles are congruent. This is true of all parallelograms.

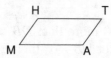

The correct choice is **(4)**.

14. Tangents to a circle from a common point are congruent, so $TH = TR$ and $RA = AP$. R is the midpoint of $\overline{TA}$, which makes $TR = RA$. Combining these congruence statements yields:

$$TH = TR = RA = AP = 10$$

In the bottom half of the figure, we can combine the given statement $\overline{HO} \cong \overline{PE}$ with the congruent tangents $HO = OZ$ and $ZE = PE$ to find:

$$HO = OZ = ZE = PE = 6$$

The perimeter is the sum of all the side lengths.

$$\text{Perimeter} = 10 + 10 + 10 + 10 + 6 + 6 + 6 + 6$$

$$= 64$$

The correct choice is **(2)**.

15. We can calculate the coordinates of the point (x, y) that divides a segment in a given ratio using the formula:

$$\frac{x - x_1}{x_2 - x} = \text{ratio and } \frac{y - y_1}{y_2 - y} = \text{ratio}$$

The coordinates of $A(-8, 7)$ are (x_1, x_2), the coordinates of $C(7, -13)$ are (y_1, y_2), and the ratio is $\frac{3}{2}$.

$$\frac{x - (-8)}{7 - x} = \frac{3}{2}$$
$$2(x + 8) = 3(7 - x)$$
$$2x + 16 = 21 - 3x$$
$$5x = 5$$
$$x = 1$$
$$\frac{y - 7}{-13 - y} = \frac{3}{2}$$
$$2(y - 7) = 3(-13 - y)$$
$$2y - 14 = -39 - 3y$$
$$5y = -25$$
$$y = -5$$

The coordinates of B are $(1, -5)$.

The correct choice is (**1**).

16. Parallel segment $\overline{DE}$ divides $\triangle ABC$ into similar triangles $\triangle ABC$ and $\triangle DBE$. Corresponding segments of similar triangles are proportional.

$$\frac{AB}{DB} = \frac{AC}{DE}$$

The given ratio $AD:DB = 3:5$ can be used to find the ratio $AB:DB$. If AD is 3 parts and DB is 5 parts, then AD is the sum of these, or 8 parts. The ratio $AB:DB$ is therefore 8:5. Now use the proportion to solve for DE.

$$\frac{8}{5} = \frac{9.4}{DE}$$
$$8(DE) = 5(9.4)$$
$$8(DE) = 47$$
$$DE = 5.875$$
$$DE = 5.9, \text{ rounded to the nearest tenth}$$

The correct choice is (**3**).

17. A reflection over a line of symmetry would map a rectangle onto itself. Using the figure, sketch the two lines of symmetry. Their equations are $y = -x + 10$ and $y = x$. Choices (1) and (2) represent reflections over these lines. A 180° rotation about the center of the rectangle will also map it onto itself. The center point $P(5, 5)$ is the point where the two lines of symmetry intersect. This rotation is represented by choice (4). Choice (3) will not map the rectangle onto itself.

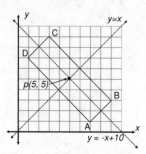

The correct choice is (3).

18. The radius of the circle is half the diameter, or 5 cm. Substitute the radius and 30° central angle into the formula for the area of a sector.

$$\text{Area of Sector} = \frac{\theta}{360}\pi r^2$$
$$= \frac{30}{360}\pi (5)^2$$
$$= 6.544$$

Rounded to the nearest tenth, the area of the sector is 6.5 cm².

The correct choice is (2).

19. First, find the area of the square base using the formula for the area of a square.

$$B = (\text{side})^2$$
$$= (60 \text{ in})^2$$
$$= 3,600 \text{ in}^2$$

Next, apply the formula for the volume of a pyramid, using the area of the square base for B and 84 inches for the height.

$$V = \frac{1}{3}Bh$$

$$= \frac{1}{3}\left(3600 \text{ in}^2\right)(84 \text{ in})$$

$$= 100,800 \text{ in}^3$$

Now convert cubic inches to cubic feet using the ratio $\dfrac{1 \text{ ft}}{12 \text{ in}}$. Remember to cube the conversion ratio since we are working with cubic feet and cubic inches.

$$V = \left(100,800 \text{ in}^3\right)\left(\frac{1 \text{ ft}}{12 \text{ ft}}\right)^3$$

$$= 58.333 \text{ ft}^3$$

Rounded to the nearest cubic foot, the volume is 58 ft³.

The correct choice is **(2)**.

20. An altitude to the hypotenuse of a right triangle forms three similar triangles with proportional sides. Sketch each of the three triangles (small △, medium △, large △), and label their sides to aid in evaluating which choice represents a correct proportion. Each side is either the hypotenuse (across from the right angle), the short leg, or the long leg.

- Choice (1): $\dfrac{AD}{AB} = \dfrac{BC}{AC}$ is equivalent to $\dfrac{\text{short leg(small }\triangle)}{\text{short leg(large }\triangle)}$

$= \dfrac{\text{hyp(medium }\triangle)}{\text{hyp(large }\triangle)}$. These are not corresponding parts.

- Choice (2): $\dfrac{AD}{AB} = \dfrac{AB}{AC}$ is equivalent to $\dfrac{\text{short leg(small }\triangle)}{\text{short leg(large }\triangle)}$

$= \dfrac{\text{hyp(small }\triangle)}{\text{hyp(large }\triangle)}$. This is a valid proportion.

- Choice (3): $\dfrac{BD}{BC} = \dfrac{AB}{AD}$ is equivalent to $\dfrac{\text{long leg (small } \triangle)}{\text{long leg (large } \triangle)}$

  $= \dfrac{\text{hyp (small } \triangle)}{\text{short leg (small } \triangle)}$. These are not corresponding parts.

- Choice (4): $\dfrac{AB}{BC} = \dfrac{BD}{AC}$ is equivalent to $\dfrac{\text{hyp (small } \triangle)}{\text{hyp (medium } \triangle)}$

  $= \dfrac{\text{long leg (small } \triangle)}{\text{hyp (large } \triangle)}$. These are not corresponding parts.

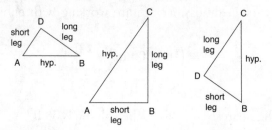

The correct choice is **(2)**.

21. The center-radius equation of a circle is $(x - h)^2 + (y - k)^2 = r^2$, where the center is (h, k) and the radius is r. Rewrite the equation of the circle in center-radius form using the completing-the-square method.

$x^2 + y^2 + 4x - 8y = -16$

$x^2 + 4x + y^2 - 8y = -16$ Rearrange to group x and y terms

$x^2 + 4x + 4 + y^2 - 8y = -16 + 4$ Add (1/2 coefficient of $x)^2$ to each side

$x^2 + 4x + 4 + y^2 - 8y + 16 =$

           $-16 + 4 + 16$ Add (1/2 coefficient of $y)^2$ to each side

$(x + 2)^2 + (y - 4)^2 = 4$ Factor

The center of the circle is $(-2, 4)$ and the radius is 2. We can eliminate choices (1) and (2) because they do not have the correct center. The center of the circle has an x-coordinate of -2, so it is 2 units left of the y-axis. The radius is 2, so the circle just reaches the y-axis and is tangent to it. You can also graph the circle to see that it is tangent to the y-axis.

The correct choice is **(4)**.

22. Perpendicular bisector $\overline{BD}$ is shown in $\triangle ABC$. D is a midpoint, which makes $\overline{BD}$ a median, so statement I is true. $\angle ADB$ and $\angle CDB$ are congruent right angles and $\overline{BD}$ is congruent to itself, so $\triangle ADB \cong \triangle CDB$ by the SAS postulate. This leads us to $\angle ABD \cong \angle CBD$ by CPCTC, making $\overline{BD}$ an angle bisector. Statement II is also true. We can also conclude that $\overline{AB} \cong \overline{BC}$ by CPCTC, which makes $\triangle ABC$ isosceles. Statement III is also true.

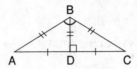

The correct choice is (**4**).

23. Dilations always create similar figures. Perimeters of similar figures are proportional to the scale factor, and areas are proportional to the scale factor squared. Given a scale factor of 3, the perimeter of the dilated image must be 3 times greater and the area is 9 times greater.

Perimeter of $R'\,J'\,M' = 12 \cdot 3 = 36$

Area of $R'\,J'\,M' = 6 \cdot 3^2 = 54$

The correct choice is (**3**).

24. The trigonometric cofunction relationship states that if $\sin(A) = \cos(B)$, then $A + B = 90°$. Given $\sin(2x + 7)° = \cos(4x - 7)°$, we can conclude:

$$2x + 7 + 4x - 7 = 90$$
$$6x = 90$$
$$x = 15$$

The correct choice is (**2**).

PART II

25. Given any chord in a circle, the perpendicular bisector of the chord passes through the center of the circle. Construct a diameter by constructing the perpendicular bisector of $\overline{AB}$:

   - With compass point at A, open the compass more than half the length of $\overline{AB}$.

   - Make arc on both sides of $\overline{AB}$.

   - Move the compass point to B.

   - With the same compass opening, make a second pair of arcs that intersect the first pair.

   - Connect the two arc intersections.

   - $\overline{MN}$ is a diameter of the circle.

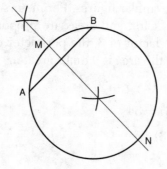

26. Starting with given $\angle A = 68°$, we know that opposite angles of a parallelogram are congruent, so $\angle BCD = 68°$. Bisector $\overline{CE}$ divides $\angle BCD$ in half, so $m\angle BCE = 34°$. Next, we know that consecutive angles of a parallelogram are supplementary, so $\angle ABC$ and $\angle A$ sum to $180°$.

$$m\angle ABC + m\angle A = 180°$$
$$m\angle ABC = 180° - m\angle A$$
$$= 180° - 68°$$
$$= 112°$$

Bisector $\overline{BE}$ divides $\angle ABC$ in half, so $m\angle EBC = 56°$. Now we can apply the angle sum theorem in $\triangle BEC$.

$$m\angle EBC + m\angle BCE + m\angle BEC = 180°$$
$$56° + 34° + m\angle BEC = 180°$$
$$90 + m\angle BEC = 180°$$
$$m\angle BEC = 90°$$

27. To find $m\angle CFE$, we need arcs $\overarc{CE}$ and $\overarc{DB}$. Calculate $\overarc{CE}$ using semicircle $\overarc{DCE}$, which we know measures $180°$:

$$m\overarc{CD} + m\overarc{CE} = 180°$$
$$46° + m\overarc{CE} = 180°$$
$$m\overarc{CE} = 134°$$

∠*CFE* is an angle formed by intersecting chords and is equal in measure to half the sum of the intercepted arcs.

$$m\angle CFE = \frac{1}{2}\left(\widehat{CE} + \widehat{DB}\right)$$

$$= \frac{1}{2}(134 + 102)$$

$$= 118°$$

28. There are a many different transformation sequences that would map *ABCD* to *A″ B″ C″ D″*. One possible sequence is a reflection over the line $y = 2$ followed by a translation 6 units down.

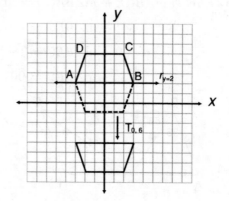

29. The stake and telephone pole are parallel. The line along the ground is a transversal that forms congruent corresponding angles. The wire is another transversal forming another pair of congruent corresponding angles. The two triangles are similar by the AA postulate.

30. Begin by rearranging the line in point-slope form so we can graph it.

$$4x + 3y = 24$$
$$3y = -4x + 24$$
$$y = -\frac{4}{3}x + 8$$

Now graph the line and the center of dilation (3, 4). From the graph, we see that the line passes through the center of dilation. This indicates that the dilated line is identical to the original line, so Aliyah is not correct.

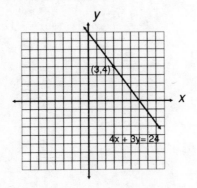

31. Each section of concrete is a rectangular prism 36 inches long, 36 inches wide, and 4 inches high. Since the cost of the concrete is given in dollars per cubic foot, convert the dimensions to feet.

$$36 \text{ in} \times \frac{1 \text{ ft}}{12 \text{ in}} = 3 \text{ ft} \qquad 4 \text{ in} \times \frac{1 \text{ ft}}{12 \text{ in}} = \frac{1}{3} \text{ ft}$$

Next, calculate the volume using the volume formula for a rectangular prism.

$$V = \text{length} \times \text{width} \times \text{height}$$
$$= (3 \text{ ft})(3 \text{ ft})\left(\frac{1}{3} \text{ ft}\right)$$
$$= 3 \text{ ft}^3$$

There are two sections of concrete, so the total volume is 6 ft³. Finally, calculate the cost.

$$\text{cost} = \text{volume} \times \text{cost/ft}^3$$
$$= \left(6 \text{ ft}^3\right)\left(\$3.25 \text{ / ft}^3\right)$$
$$= \$19.50$$

PART III

32. The strategy for this proof is to use the theorem that all points on the perpendicular bisector of a segment are equidistant from the endpoints of the segment. Start by proving $\triangle ABD \cong \triangle CBD$. Then $\overline{AD} \cong \overline{CD}$ by CPCTC. We now have $\overline{AD} \cong \overline{CD}$ and $\overline{AB} \cong \overline{CB}$, which indicate that points B and D are equidistant from endpoints A and C. This is sufficient to prove $\overline{BDE}$ is the perpendicular bisector of $\overline{AC}$.

Statements	Reasons
(1) $\triangle ABC$, $\overline{AEC}$, $\overline{BDE}$ with $\angle ABE \cong \angle CBE$ and $\angle ADE \cong \angle CDE$	(1) Given
(2) $\overline{BD} \cong \overline{BD}$	(2) Reflexive Property
(3) $\angle BDA$ and $\angle ADE$ are supplementary. $\angle BDC$ and $\angle CDE$ are supplementary.	(3) Linear pairs of angles are supplementary.
(4) $\angle BDA \cong \angle BDC$	(4) Supplements of congruent angles are congruent.
(5) $\triangle ABD \cong \triangle CBD$	(5) ASA
(6) $\overline{AD} \cong \overline{CD}$, $\overline{AB} \cong \overline{CB}$	(6) CPCTC
(7) $\overline{BDE}$ is the perpendicular bisector of $\overline{AC}$.	(7) The perpendicular bisector of a segment is the set of points equidistant from the endpoints of the segment.

33. Focusing just on $\triangle AHG$ and segment $\overline{AB}$, we have transversal $\overline{AG}$ intersecting parallel segments $\overline{HG}$ and $\overline{AB}$. $\angle AHG$ and $\angle BAG$ are congruent alternate interior angles, so m$\angle G = 36°$. We can use a trigonometric ratio to find the rise HA. Relative to $\angle G$, $\overline{HA}$ is the opposite and $\overline{HG}$ is the adjacent, so we use the tangent ratio.

$$\tan G = \frac{opposite}{adjacent}$$

$$\tan G = \frac{HA}{HG}$$

$$\tan 36 = \frac{HA}{10}$$

$$HA = 7.265$$

Rounded to the nearest tenth of an inch, the run is 7.3 inches.

To find AC, we can calculate AG and multiply by 3. AG is a hypotenuse relative to $\angle G$, so we can use a cosine ratio:

$$\cos G = \frac{adjacent}{hypotenuse}$$

$$\cos G = \frac{HG}{AG}$$

$$\cos 36 = \frac{10}{AG}$$

$$AG = 12.360$$

Now calculate AC:

$$AC = 3AG$$
$$= 3(12.36)$$
$$= 37.082$$

Rounded to the nearest inch, $AC = 37$ inches.

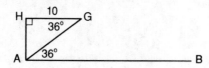

34. Each chocolate sphere is actually a spherical shell with an inner and outer radius. The outer radius is half the diameter, or 2 cm. The inner radius is found by subtracting the wall thickness from the outer radius.

$$r_{in} = r_{out} - \text{thickness}$$
$$= 2\,\text{cm} - 0.5\,\text{cm}$$
$$= 1.5\,\text{cm}$$

Now use the formula for the volume of a sphere to find each volume.

$$V_{out} = \frac{4}{3}\pi r^3$$
$$= \frac{4}{3}\pi(2)^3$$
$$= 33.51032\,\text{cm}^3$$
$$V_{in} = \frac{4}{3}\pi r^3$$
$$= \frac{4}{3}\pi(1.5)^3$$
$$= 14.13716\,\text{cm}^3$$

The volume of chocolate is equal to the difference between these two volumes.

$$V_{chocolate} = V_{out} - V_{in}$$
$$= 33.51032\,\text{cm}^3 - 14.13716\,\text{cm}^3$$
$$= 19.37316$$

Rounded to the nearest tenth of a cubic centimeter, the volume is 19.4 cm^3.

Use the relationship *mass = volume × density* to find the mass of 1 chocolate, and multiply by 8 because there are 8 chocolates per box.

$$\text{mass} = 8 \times \text{volume} \times \text{density}$$
$$= 8(19.4\,\text{cm}^3)(1.308\text{g/cm}^3)$$
$$= 203.0016\,\text{g}$$

To the nearest gram, the mass of chocolate in one box is 203 g.

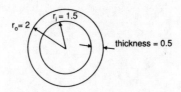

PART IV

35. There are many different combinations of properties we can use to prove that a figure is both a parallelogram and rectangle. One option is to find the slopes of the four sides, which can be used to show that the opposite sides and parallel and consecutive sides are perpendicular.

a) Calculate each slope using coordinates $M(-4, 2)$, $A(-1, -3)$, $T(9, 3)$, and $H(6, 8)$.

$$\text{slope} = \frac{y_2 - y_1}{x_2 - x_1}$$

$$\text{slope of } \overline{MA} : \frac{-3 - 2}{-1 - (-4)} = -\frac{5}{3}$$

$$\text{slope of } \overline{AT} : \frac{3 - (-3)}{9 - (-1)} = \frac{6}{10} = \frac{3}{5}$$

$$\text{slope of } \overline{TH} : \frac{8 - 3}{6 - 9} = -\frac{5}{3}$$

$$\text{slope of } \overline{HM} : \frac{8 - 2}{6 - (-4)} = \frac{6}{10} = \frac{3}{5}$$

Opposite sides of *MATH* have equal slopes, which makes them parallel. A quadrilateral with opposite sides parallel is a parallelogram, therefore *MATH* is a parallelogram.

b) The slopes of consecutive sides are negative reciprocals, which makes them perpendicular. A parallelogram with perpendicular consecutive sides is a rectangle.

Topic	Question Numbers	Number of Points	Your Points	Your Percentage
1. Basic Angle and Segment Relationships	1	2		
2. Angle and Segment Relationships in Triangles and Polygons	22	2		
3. Constructions	25	2		
4. Transformations	2, 4, 17, 28	$2 + 2 + 2 + 2 = 8$		
5. Triangle Congruence	10, 32	$2 + 4 = 6$		
6. Lines, Segments, and Circles on the Coordinate Plane	8, 11, 15, 21, 30	$2 + 2 + 2 + 2 + 2 = 10$		
7. Similarity	7, 12, 16, 20, 23, 29	$2 + 2 + 2 + 2 + 2 + 2 = 12$		
8. Trigonometry	6, 9, 24, 33	$2 + 2 + 2 + 4 = 10$		
9. Parallelograms	13, 26	$2 + 2 = 4$		
10. Coordinate Geometry Proofs	35	6		
11. Volume, Solids, and Modeling	3, 5, 19, 31, 34	$2 + 2 + 2 + 2 + 4 = 12$		
12. Circles	14, 18, 27	$2 + 2 + 2 = 6$		

Regents Examination in Geometry—August 2018
Chart for Converting Total Test Raw Scores to Final
Examination Scores (Scaled Scores)

Raw Score	Scaled Score	Performance Level	Raw Score	Scaled Score	Performance Level	Raw Score	Scaled Score	Performance Level
80	100	5	53	80	4	26	61	2
79	99	5	52	80	4	25	60	2
78	98	5	51	79	3	24	58	2
77	97	5	50	79	3	23	57	2
76	96	5	49	78	3	22	56	2
75	95	5	48	78	3	21	55	2
74	95	5	47	77	3	20	53	1
73	94	5	46	77	3	19	51	1
72	93	5	45	76	3	18	50	1
71	92	5	44	76	3	17	48	1
70	91	5	43	75	3	16	46	1
69	91	5	42	74	3	15	44	1
68	90	5	41	74	3	14	42	1
67	89	5	40	73	3	13	40	1
66	88	5	39	73	3	12	38	1
65	88	5	38	72	3	11	36	1
64	87	5	37	71	3	10	33	1
63	86	5	36	70	3	9	31	1
62	86	5	35	70	3	8	28	1
61	85	5	34	69	3	7	26	1
60	84	4	33	68	3	6	23	1
59	84	4	32	67	3	5	19	1
58	83	4	31	66	3	4	16	1
57	83	4	30	65	3	3	3	1
56	82	4	29	64	2	2	9	1
55	81	4	28	63	2	1	5	1
54	81	4	27	62	2	0	0	1

Examination
June 2019
Geometry

HIGH SCHOOL MATH REFERENCE SHEET

Conversions

1 inch = 2.54 centimeters

1 meter = 39.37 inches

1 mile = 5280 feet

1 mile = 1760 yards

1 mile = 1.609 kilometers

1 kilometer = 0.62 mile

1 pound = 16 ounces

1 pound = 0.454 kilogram

1 kilogram = 2.2 pounds

1 ton = 2000 pounds

1 cup = 8 fluid ounces

1 pint = 2 cups

1 quart = 2 pints

1 gallon = 4 quarts

1 gallon = 3.785 liters

1 liter = 0.264 gallon

1 liter = 1000 cubic centimeters

Formulas

Triangle	$A = \frac{1}{2}bh$
Parallelogram	$A = bh$
Circle	$A = \pi r^2$
Circle	$C = \pi d$ or $C = 2\pi r$

Formulas (continued)

General Prisms	$V = Bh$
Cylinder	$V = \pi r^2 h$
Sphere	$V = \frac{4}{3}\pi r^3$
Cone	$V = \frac{1}{3}\pi r^2 h$
Pyramid	$V = \frac{1}{3}Bh$
Pythagorean Theorem	$a^2 + b^2 = c^2$
Quadratic Formula	$x = \dfrac{-b \pm \sqrt{b^2 - 4ac}}{2a}$
Arithmetic Sequence	$a_n = a_1 + (n-1)d$
Geometric Sequence	$a_n = a_1 r^{n-1}$
Geometric Series	$S_n = \dfrac{a_1 - a_1 r^n}{1 - r}$ where $r \neq 1$
Radians	1 radian $= \dfrac{180}{\pi}$ degrees
Degrees	1 degree $= \dfrac{\pi}{180}$ radians
Exponential Growth/Decay	$A = A_0 e^{k(t - t_0)} + B_0$

180 Minutes–35 Questions

PART I

Answer all 24 questions in this part. Each correct answer will receive 2 credits. No partial credit will be allowed. pick up after "allowed." with the following: For each statement or question, write in the space provided the numeral preceding the word or expression that best completes the statement or answers the question.   [48 credits]

1 On the set of axes below, triangle ABC is graphed. Triangles $A'B'C'$ and $A''B''C''$, the images of triangle ABC, are graphed after a sequence of rigid motions.

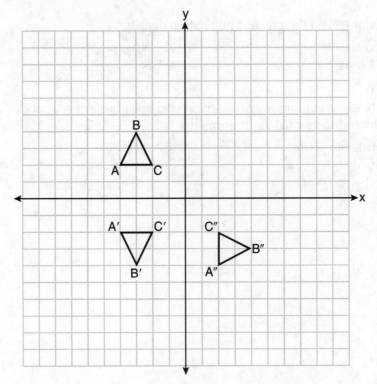

Identify which sequence of rigid motions maps $\triangle ABC$ onto $\triangle A'B'C'$ and then maps $\triangle A'B'C'$ onto $\triangle A''B''C''$.

(1) a rotation followed by another rotation
(2) a translation followed by a reflection
(3) a reflection followed by a translation
(4) a reflection followed by a rotation   1 _____

2 The table below shows the population and land area, in square miles, of four counties in New York State at the turn of the century.

County	2000 Census Population	2000 Land Area (mi^2)
Broome	200,536	706.82
Dutchess	280,150	801.59
Niagara	219,846	522.95
Saratoga	200,635	811.84

Which county had the greatest population density?

(1) Broome   (3) Niagara

(2) Dutchess   (4) Saratoga   2 _____

3 If a rectangle is continuously rotated around one of its sides, what is the three-dimensional figure formed?

(1) rectangular prism   (3) sphere

(2) cylinder   (4) cone   3 _____

4 Which transformation carries the parallelogram below onto itself?

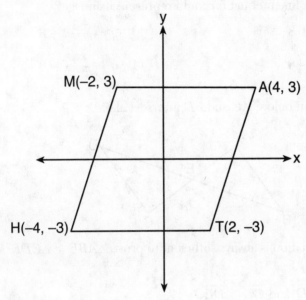

(1) a reflection over $y = x$

(2) a reflection over $y = -x$

(3) a rotation of 90° counterclockwise about the origin

(4) a rotation of 180° counterclockwise about the origin　　4 ____

5 After a dilation centered at the origin, the image of $\overline{CD}$ is $\overline{C'D'}$. If the coordinates of the endpoints of these segments are $C(6, -4)$, $D(2, -8)$, $C'(9, -6)$, and $D'(3, -12)$, the scale factor of the dilation is

(1) $\dfrac{3}{2}$　　　　　　　　　　　　(3) 3

(2) $\dfrac{2}{3}$　　　　　　　　　　　　(4) $\dfrac{1}{3}$　　5 ____

6 A tent is in the shape of a right pyramid with a square floor. The square floor has side lengths of 8 feet. If the height of the tent at its center is 6 feet, what is the volume of the tent, in cubic feet?

(1) 48　　　　　　　　　　　　(3) 192

(2) 128　　　　　　　　　　　　(4) 384　　6 ____

7 The line $-3x + 4y = 8$ is transformed by a dilation centered at the origin. Which linear equation could represent its image?

(1) $y = \frac{4}{3}x + 8$ (3) $y = -\frac{3}{4}x - 8$

(2) $y = \frac{3}{4}x + 8$ (4) $y = -\frac{4}{3}x - 8$ 7_____

8 In the diagram below, $\overline{AC}$ and $\overline{BD}$ intersect at E.

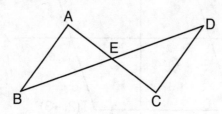

Which information is always sufficient to prove $\triangle ABE \cong \triangle CDE$?

(1) $\overline{AB} \parallel \overline{CD}$

(2) $\overline{AB} \cong \overline{CD}$ and $\overline{BE} \cong \overline{DE}$

(3) E is the midpoint of $\overline{AC}$.

(4) $\overline{BD}$ and $\overline{AC}$ bisect each other.   8_____

9 The expression $\sin 57°$ is equal to

(1) $\tan 33°$ (3) $\tan 57°$

(2) $\cos 33°$ (4) $\cos 57°$ 9_____

10 What is the volume of a hemisphere that has a diameter of 12.6 cm, to the *nearest tenth of a cubic centimeter*?

(1) 523.7   (3) 4189.6

(2) 1047.4   (4) 8379.2   10_____

11 In the diagram below of △ABC, D is a point on $\overline{BA}$, E is a point on $\overline{BC}$, and $\overline{DE}$ is drawn.

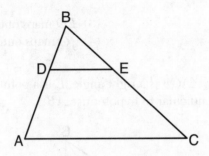

If $BD = 5$, $DA = 12$, and $BE = 7$, what is the length of $\overline{BC}$ so that $\overline{AC} \parallel \overline{DE}$?

(1) 23.8   (3) 15.6

(2) 16.8   (4) 8.6   11 _____

12 A quadrilateral must be a parallelogram if

(1) one pair of sides is parallel and one pair of angles is congruent

(2) one pair of sides is congruent and one pair of angles is congruent

(3) one pair of sides is both parallel and congruent

(4) the diagonals are congruent   12 _____

13 In the diagram below of circle O, chords $\overline{JT}$ and $\overline{ER}$ intersect at M.

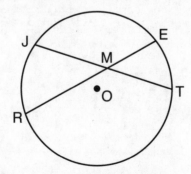

If $EM = 8$ and $RM = 15$, the lengths of $\overline{JM}$ and $\overline{TM}$ could be

(1) 12 and 9.5   (3) 16 and 7.5

(2) 14 and 8.5   (4) 18 and 6.5   13 _____

14 Triangles *JOE* and *SAM* are drawn such that $\angle E \cong \angle M$ and $\overline{EJ} \cong \overline{MS}$. Which mapping would *not* always lead to $\triangle JOE \cong \triangle SAM$?

(1) $\angle J$ maps onto $\angle S$ (3) $\overline{EO}$ maps onto $\overline{MA}$

(2) $\angle O$ maps onto $\angle A$ (4) $\overline{JO}$ maps onto $\overline{SA}$ 14 _____

15 In $\triangle ABC$ shown below, $\angle ACB$ is a right angle, E is a point on $\overline{AC}$, and $\overline{ED}$ is drawn perpendicular to hypotenuse $\overline{AB}$.

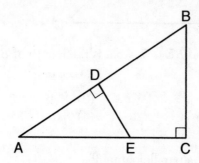

If $AB = 9$, $BC = 6$, and $DE = 4$, what is the length of $\overline{AE}$?

(1) 5   (3) 7

(2) 6   (4) 8   15 _____

16 Which equation represents a line parallel to the line whose equation is $-2x + 3y = -4$ and passes through the point $(1, 3)$?

(1) $y - 3 = -\frac{3}{2}(x - 1)$ (3) $y + 3 = -\frac{3}{2}(x + 1)$

(2) $y - 3 = \frac{2}{3}(x - 1)$ (4) $y + 3 = \frac{2}{3}(x + 1)$ 16 _____

17 In rhombus *TIGE*, diagonals $\overline{TG}$ and $\overline{IE}$ intersect at *R*. The perimeter of *TIGE* is 68, and *TG* = 16.

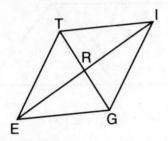

What is the length of diagonal $\overline{IE}$?

(1) 15   (3) 34

(2) 30   (4) 52   17 _____

18 In circle *O* two secants, $\overline{ABP}$ and $\overline{CDP}$, are drawn to external point *P*. If m$\widehat{AC}$ = 72°, and m$\widehat{BD}$ = 34°, what is the measure of $\angle P$?

(1) 19°   (3) 53°

(2) 38°   (4) 106°   18 _____

19 What are the coordinates of point *C* on the directed segment from $A(-8, 4)$ to $B(10, -2)$ that partitions the segment such that *AC:CB* is 2:1?

(1) (1, 1)   (3) (2, −2)

(2) (−2, 2)   (4) (4, 0)   19 _____

20 The equation of a circle is $x^2 + 8x + y^2 - 12y = 144$. What are the coordinates of the center and the length of the radius of the circle?

(1) center (4, −6) and radius 12

(2) center (−4, 6) and radius 12

(3) center (4, −6) and radius 14

(4) center (−4, 6) and radius 14   20 _____

21 In parallelogram $PQRS$, $\overline{QP}$ is extended to point T and $\overline{ST}$ is drawn.

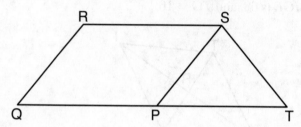

If $\overline{ST} \cong \overline{SP}$ and m$\angle R = 130°$, what is m$\angle PST$?

(1) 130°   (3) 65°

(2) 80°   (4) 50°   21 _____

22 A 12-foot ladder leans against a building and reaches a window 10 feet above ground. What is the measure of the angle, to the *nearest degree*, that the ladder forms with the ground?

(1) 34   (3) 50

(2) 40   (4) 56   22 _____

23 In the diagram of equilateral triangle ABC shown below, E and F are the midpoints of $\overline{AC}$ and $\overline{BC}$, respectively.

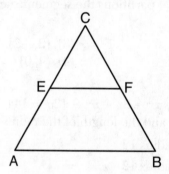

If $EF = 2x + 8$ and $AB = 7x - 2$, what is the perimeter of trapezoid $ABFE$?

(1) 36   (3) 100

(2) 60   (4) 120   23 _____

24 Which information is *not* sufficient to prove that a parallelogram is a square?

(1) The diagonals are both congruent and perpendicular.

(2) The diagonals are congruent and one pair of adjacent sides are congruent.

(3) The diagonals are perpendicular and one pair of adjacent sides are congruent.

(4) The diagonals are perpendicular and one pair of adjacent sides are perpendicular.   24 _____

PART II

Answer all 7 questions in this part. Each correct answer will receive 2 credits. Clearly indicate the necessary steps, including appropriate formula substitutions, diagrams, graphs, charts, etc. For all questions in this part, a correct numerical answer with no work shown will receive only 1 credit.   [14 credits]

25 Triangle $A'B'C'$ is the image of triangle ABC after a dilation with a scale factor of $\frac{1}{2}$ and centered at point A. Is triangle ABC congruent to triangle $A'B'C'$? Explain your answer.

26 Determine and state the area of triangle PQR, whose vertices have coordinates $P(-2, -5)$, $Q(3, 5)$, and $R(6, 1)$.

[The use of the set of axes below is optional.]

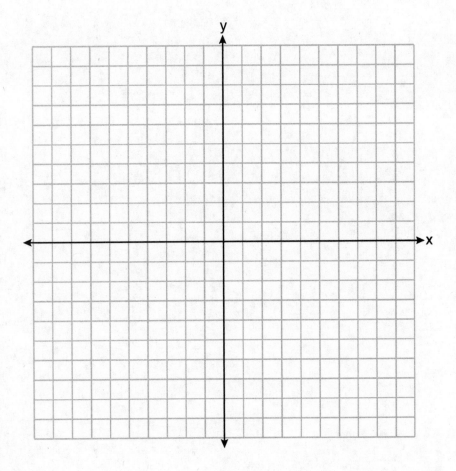

27 A support wire reaches from the top of a pole to a clamp on the
ground. The pole is perpendicular to the level ground and the
clamp is 10 feet from the base of the pole. The support wire makes
a 68° angle with the ground. Find the length of the support wire to
the *nearest foot*.

28 In the diagram below, circle O has a radius of 10.

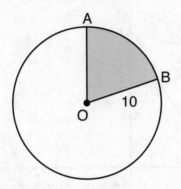

If $m\overset{\frown}{AB} = 72°$, find the area of shaded sector AOB, in terms of π.

29 On the set of axes below, $\triangle ABC \cong \triangle STU$.

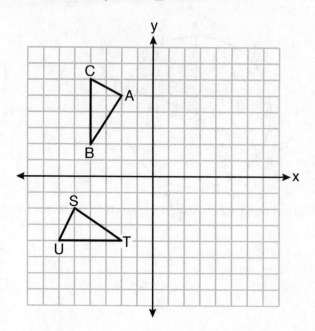

Describe a sequence of rigid motions that maps $\triangle ABC$ onto $\triangle STU$.

30 In right triangle *PRT*, m∠*P* = 90°, altitude $\overline{PQ}$ is drawn to hypotenuse $\overline{RT}$, *RT* = 17, and *PR* = 15.

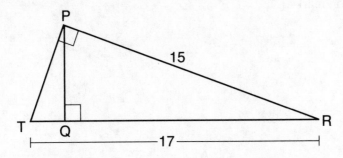

Determine and state, to the *nearest tenth*, the length of $\overline{RQ}$.

31 Given circle O with radius $\overline{OA}$, use a compass and straightedge to construct an equilateral triangle inscribed in circle O. [Leave all construction marks.]

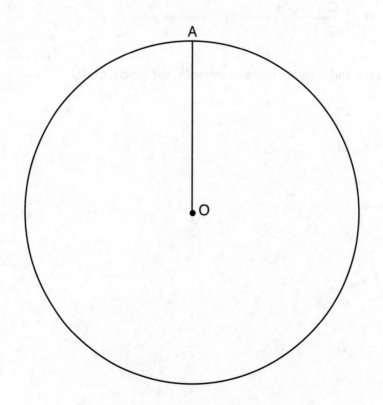

PART III

Answer all 3 questions in this part. Each correct answer will receive 4 credits. Clearly indicate the necessary steps, including appropriate formula substitutions, diagrams, graphs, charts, etc. Utilize the information provided for each question to determine your answer. Note that diagrams are not necessarily drawn to scale. For all questions in this part, a correct numerical answer with no work shown will receive only 1 credit.   [12 credits]

32   Riley plotted $A(-1, 6)$, $B(3, 8)$, $C(6, -1)$, and $D(1, 0)$ to form a quadrilateral. Prove that Riley's quadrilateral $ABCD$ is a trapezoid. [The use of the set of axes on the next page is optional.]

Question 32 is continued on the next page.

Question 32 continued

Riley defines an isosceles trapezoid as a trapezoid with congruent diagonals. Use Riley's definition to prove that *ABCD* is *not* an isosceles trapezoid.

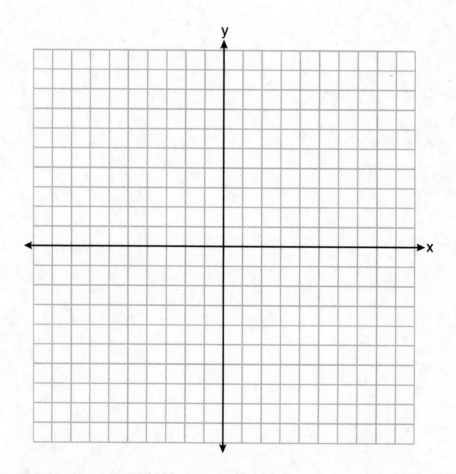

33 A child-sized swimming pool can be modeled by a cylinder. The pool has a diameter of $6\frac{1}{2}$ feet and a height of 12 inches. The pool is filled with water to $\frac{2}{3}$ of its height. Determine and state the volume of the water in the pool, to the *nearest cubic foot*.

One cubic foot equals 7.48 gallons of water. Determine and state, to the *nearest gallon*, the number of gallons of water in the pool.

34 Nick wanted to determine the length of one blade of the windmill
pictured below. He stood at a point on the ground 440 feet from
the windmill's base. Using surveyor's tools, Nick measured the angle
between the ground and the highest point reached by the top blade
and found it was 38.8°. He also measured the angle between the
ground and the lowest point of the top blade, and found it was 30°.

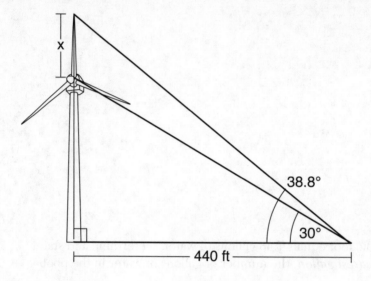

Determine and state a blade's length, x, to the *nearest foot*.

PART IV

Answer the question in this part. Each correct answer will receive 6 credits. Clearly indicate the necessary steps, including appropriate formula substitutions, diagrams, graphs, charts, etc. For all questions in this part, a correct numerical answer with no work shown will receive only 1 credit. [6 credits]

35 Given: Quadrilateral $MATH$, $\overline{HM} \cong \overline{AT}$, $\overline{HT} \cong \overline{AM}$, $\overline{HE} \perp \overline{MEA}$ and $\overline{HA} \perp \overline{AT}$

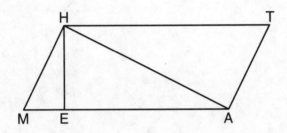

Prove: $TA \bullet HA = HE \bullet TH$

Work space for question 35 is continued on the next page.

Question 35 continued

Answers
June 2019
Geometry

Answer Key

PART I

1. 4	**5.** 1	**9.** 2	**13.** 3	**17.** 2	**21.** 2
2. 3	**6.** 2	**10.** 1	**14.** 4	**18.** 1	**22.** 4
3. 2	**7.** 2	**11.** 1	**15.** 2	**19.** 4	**23.** 3
4. 4	**8.** 4	**12.** 3	**16.** 2	**20.** 4	**24.** 3

PART II

25. The triangles are not congruent because corresponding sides are not congruent.

26. area $= 25$

27. length $= 27$ ft

28. area $= 20\pi$

29. Rotate 90° counterclockwise about the origin.

30. $RQ = 13.2$

31. See the detailed solution for the construction.

PART III

32. $ABCD$ is a trapezoid because $\overline{AD} \parallel \overline{BC}$.

ABCD is not isosceles because $\overline{AC}$ is not congruent to $\overline{BD}$.

33. volume $= 22$ ft^3, number of gallons $= 165$

34. length $= 100$ ft

PART IV

35. See the detailed solution for the proof.

In **PARTS II–IV,** you are required to show how you arrived at your answers. For sample methods of solutions, see the *Answers Explained* section.

Answers Explained

PART I

1. The sequence of transformations needs to map the following the points:

$$A(-4, 2) \rightarrow A'(-4, -2) \rightarrow A''(2, -4).$$

A reflection over the x-axis changes the sign of the y-coordinate, and takes A to A'. A counterclockwise rotation of $90°$ centered at the origin changes coordinates (x, y) to $(-y, x)$, and takes A' to A''. This same sequence also maps B to B'' and C to C''.

The correct choice is (**4**).

2. Find the population density of each county by dividing the population by the area.

Broome: population density $= \dfrac{200,536}{706.82} = 283.716$

Dutchess: population density $= \dfrac{280,150}{801.59} = 349.493$

Niagara: population density $= \dfrac{219,846}{522.95} = 428.396$

Saratoga: population density $= \dfrac{200,635}{811.84} = 247.136$

Niagara county has the greatest population density.

The correct choice is (**3**).

3. The solid formed by rotating a rectangle about one of its sides is a cylinder.

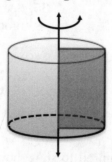

The correct choice is (**2**).

4. One approach is to choose a vertex and apply each transformation. We are looking to map the chosen vertex to one of the other vertices. If the vertex does not map to one of the original vertices, that transformation can be eliminated. Using vertex $M(-2, 3)$ we apply each transformation.

Choice 1 – A reflection over the line $y = x$ swaps the coordinates with the rule $(x, y) \rightarrow (y, x)$. $M(-2, 3) \rightarrow M'(3, -2)$ which is not a vertex, so this choice is eliminated.

Choice 2 – A reflection over the line $y = -x$ swaps the coordinates and changes the signs with the rule $(x, y) \rightarrow (-y, -x)$. $M(-2, 3) \rightarrow M'(-3, 2)$ which is not a vertex, so this choice is eliminated.

Choice 3 – A counterclockwise rotation of $90°$ about the origin transforms the coordinates with the rule $(x, y) \rightarrow (-y, x)$. $M(-2, 3) \rightarrow M'(-3, -2)$ which is not a vertex, so this choice is eliminated.

Choice 4 – A counterclockwise rotation of $180°$ about the origin transforms the coordinates with the rule $(x, y) \rightarrow (-x, -y)$. $M(-2, 3) \rightarrow M'(2, -3)$ which is vertex T. You should confirm that the same rule maps A to H, H to A, and T to M.

The correct choice is **(4)**.

5. A dilation always multiples each coordinate by the scale factor. Work backwards by dividing the new coordinate by the original to calculate the scale factor. We can use any point and either coordinate. Using the x-coordinates of points C and C' we find:

$$\text{scale factor} = \frac{x\text{-coordinate of } C'}{x\text{-coordinate of } C}$$

$$= \frac{9}{6}$$

$$= \frac{3}{2}$$

You can confirm this result using the other coordinates.

The correct choice is **(1)**.

6. The volume of a pyramid is given by $V = \frac{1}{3}Bh$ where B is the area of the base and h is the height. First find the area of the square base.

$$B = \text{side}^2$$
$$= 8^2$$
$$= 64$$

Now apply the volume formula with a height of 6.

$$V = \frac{1}{3}Bh$$
$$= \frac{1}{3}(64)(6)$$
$$= 128$$

The volume of the tent is 128 ft^2.

The correct choice is (**2**).

7. Dilating a line through the origin multiplies the y-intercept by the scale factor, but leaves the slope unchanged. The first step is to identify the slope of the original line by rewriting it in $y = mx + b$ form.

$$-3x + 4y = 8$$
$$4y = 3x + 8$$
$$y = \frac{3}{4}x + 2$$

The slope of the line is $\frac{3}{4}$ and the y-intercept is 2. A dilated line must also have a slope of $\frac{3}{4}$. The only choice with the correct slope is $y = \frac{3}{4}x + 8$.

The correct choice is (**2**).

8. The only congruent parts we are given are the vertical angles, $\angle AEB \cong \angle DEC$. Examine each of the choices and determine if it provides enough information to apply one of the triangle congruence postulates.

Choice 1 – Parallel sides $\overline{AB}$ and $\overline{CD}$ result in congruent alternate interior angles, so we can conclude $\angle A \cong \angle C$ and $\angle B \cong \angle D$. All three angles are congruent, but AAA is not a congruence postulate.

Choice 2 – $\overline{AB} \cong \overline{CD}$ and $\overline{BE} \cong \overline{DE}$ give 2 pairs of congruent sides, but the congruent angles are not the included angles. This choice results in ASS which is not a congruence postulate.

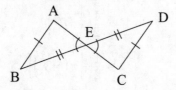

Choice 3 – Midpoint E lets us conclude $\overline{AE} \cong \overline{AC}$. This one additional pair of sides results in SA, which is not a congruence postulate.

Choice 4 – $\overline{BD}$ and $\overline{AC}$ bisecting each other implies E is the midpoint of $\overline{AC}$ and $\overline{BD}$.

Therefore, $\overline{BE} \cong \overline{ED}$ and $\overline{AE} \cong \overline{EC}$ for SAS, which is one of the triangle congruence postulates.

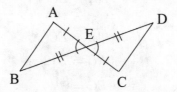

The correct choice is **(4)**.

9. The co-function relationship states that if two angles A and B sum to $90°$, then $\sin(A) = \cos(B)$. Since $57° + 33° = 90°$, $\sin(57°)$ must equal $\cos(33°)$. Alternatively, you can use your calculator to evaluate each of the expressions. Both $\sin(57°)$ and $\cos(33°)$ are equal to 0.838671.

The correct choice is **(2)**.

10. The radius of the hemisphere is found by dividing the 12.6 cm diameter by 2, so the radius is 6.3 cm. A hemisphere is half of a sphere, so multiply the volume formula of a sphere by $\frac{1}{2}$.

$$V = \frac{1}{2}\left(\frac{4}{3}\pi R^3\right)$$

$$= \frac{1}{2}\left(\frac{4}{3}\pi\left(6.3^3\right)\right)$$

$$= 523.7$$

The correct choice is (**1**).

11. A segment parallel to a side of a triangle forms a triangle similar to the original. $\triangle BDE \sim \triangle BAC$, and we can use the side splitter theorem to solve for EC.

$$\frac{BD}{DA} = \frac{BE}{EC}$$

$$\frac{5}{12} = \frac{7}{EC}$$

$$5EC = 7 \cdot 12$$

$$EC = \frac{84}{5}$$

$$= 16.8$$

Now calculate BC.

$$BC = BE + EC$$

$$= 7 + 16.8$$

$$= 23.8$$

The correct choice is (**1**).

12. The only choice that always specifies a parallelogram is choice (3). A counter-example can be identified for each of the other choices as shown in the figures.

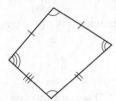

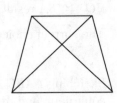

1 pair of parallel sides   1 pair of congruent sides

1 pair of congruent angles   1 pair of congruent angles   congruent diagonals

The correct choice is (3).

13. When two chords intersect in a circle, the products of the parts of the chords are equal.

$$JM \cdot TM = EM \cdot RM$$
$$JM \cdot TM = 8 \cdot 15$$
$$JM \cdot TM = 120$$

Now check which pair has a product of 120. The only choice that works is $16 \cdot 7.5 = 120$.

The correct choice is (3).

14. Sketch the two triangles with the given congruent parts marked. Check each choice to determine which additional pair of parts would not result in a valid triangle congruence postulate.

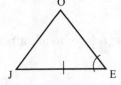

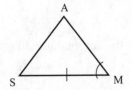

$\angle J \cong \angle S$ results in triangles congruent by ASA.

$\angle O \cong \angle A$ results in triangles congruent by AAS.

$\overline{EO} \cong \overline{MA}$ results in triangles congruent by SAS.

$\overline{JO} \cong \overline{SA}$ results in ASS, which is not a triangles congruence postulate.

The correct choice is **(4)**.

15. Sketching $\triangle ABC$ and $\triangle AED$ separately, we see that right angles C and D are congruent, and the shared angle A is also congruent. Therefore, the triangles are similar by the AA postulate. The similarity statement is $\triangle ABC \sim \triangle AED$. Corresponding sides of similar triangles are proportional, so we can set up a proportion and solve for AE.

$$\frac{AE}{AB} = \frac{ED}{BC}$$

$$\frac{AE}{9} = \frac{4}{6}$$

$$6AE = 9 \cdot 4$$

$$AE = 6$$

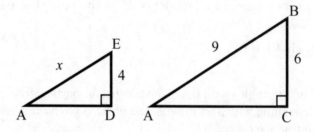

The correct choice is **(2)**.

16. First rewrite the original line in $y = mx + b$ form to identify the slope.

$$-2x + 3y = -4$$

$$3y = 2x - 4$$

$$y = \frac{2}{3}x - \frac{4}{3}$$

The slope of the original line is $\frac{2}{3}$. Parallel lines have equal slopes, so the new line must also have a slope of $\frac{2}{3}$. Write the equation of the parallel line using the point-slope form $y - y_1 = m(x - x_1)$.

Substitute $\frac{2}{3}$ for m and the given coordinates (1, 3) for x_1 and y_1 to get

$$y - 3 = \frac{2}{3}(x - 1).$$

The correct choice is **(2)**.

17. All sides of a rhombus are congruent, so divide the perimeter by 4 to find each side length.

$$\text{side length} = \frac{1}{4} \text{ perimeter}$$
$$= \frac{68}{4}$$
$$= 17$$

The diagonals of a rhombus bisect each other, so divide TG by 2 to find TR and RG

$$TR = RG = \frac{16}{2}$$
$$TR = RG = 8$$

The diagonals of a rhombus are perpendicular, making $\triangle GRE$ a right triangle. Apply the Pythagorean theorem in $\triangle GRE$ to find RE.

$$a^2 + b^2 = c^2$$
$$RE^2 + RG^2 = EG^2$$
$$RE^2 + 8^2 = 17^2$$
$$RE^2 + 64 = 289$$
$$RE^2 = 225$$
$$RE = \sqrt{225}$$
$$= 15$$

Diagonal $\overline{IE}$ is twice the length of $\overline{RE}$, so $IE = 30$.

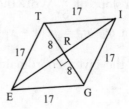

The correct choice is (**2**).

18. The angle formed by two secants from an external point is equal to half the difference of the intercepted arcs.

$$m\angle P = \frac{1}{2}\left(m\widehat{AC} - m\widehat{BD}\right)$$

$$= \frac{1}{2}(72 - 34)$$

$$= 19$$

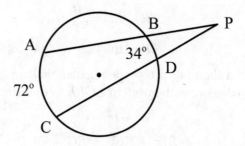

The correct choice is (**1**).

19. There are several different methods for solving a directed line segment problem. A graphical approach is to sketch the starting and ending x-coordinates on a number line. Then using the given ratio, counts jumps from the left and right. The two ends will meet at the desired x-coordinate. In this case the ratio is 2:1, so count two jumps from the left and 1 jump from the right.

After 6 jumps the two ends meet at an x-coordinate of 4. The only choice with an x-coordinate of 4 is $(4, 0)$. If needed, you can repeat this procedure for the y-coordinate.

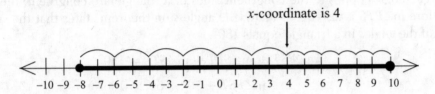

x-coordinate is 4

The correct choice is **(4)**.

20. Use the completing the square procedure to rewrite the equation in the form $(x - h)^2 + (y - k)^2 = r^2$.

First find the constant terms needed to complete the square:

$$\text{constant for the } x\text{-terms} = (\tfrac{1}{2} \cdot \text{coefficient of } x)^2$$
$$= \left(\frac{1}{2} \cdot 8\right)^2$$
$$= (4)^2$$
$$= 16$$

The constant for the y-terms $= (\tfrac{1}{2} \cdot \text{coefficient of } y)^2$
$$= \left(\frac{1}{2} \cdot (-12)\right)^2$$
$$= (-6)^2$$
$$= 36$$

Add the required constant to each side of the equation and factor.

$$x^2 + 8x + 16 + y^2 - 12y + 36 = 144 + 16 + 36$$
$$(x + 4)^2 + (y - 6)^2 = 196$$

The values of h and k are -4 and 6. Be careful with the signs because h and k are the values subtracted from x and y. The center is located at $(-4, 6)$. Since $r^2 = 196$, the radius is equal to $\sqrt{196}$, or 14.

The correct choice is **(4)**.

21. Starting with m∠R = 130°, we know that m∠QPS = 130° because opposite angles of a parallelogram are congruent. ∠QPS and ∠SPT are a linear pair which sum to 180°, making m∠SPT = 50°. △SPT is isosceles with ST = SP. The angles opposite the congruent sides of a triangle are congruent, therefore m∠PTS is also 50°. Finally, the angle sum theorem states that the sum of the angles in a triangle equals 180°.

$$m\angle PST + m\angle SPT + m\angle STP = 180°$$

$$m\angle PST + 50° + 50° = 180°$$

$$m\angle PST = 80°$$

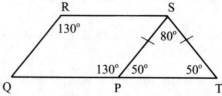

The correct choice is **(2)**.

22. The ladder, building and ground form a right triangle as shown. Trigonometry can be used to find the unknown angle using the two known sides. Relative to the unknown angle, the 12 foot ladder is the hypotenuse and the 10 foot height is the opposite. The opposite and hypotenuse indicate that a sine ratio is needed. After setting up the ratio, apply the inverse sine function to find the angle.

$$\sin(x) = \frac{\text{opposite}}{\text{hypotenuse}}$$

$$\sin(x) = \frac{10}{12}$$

$$x = \sin^{-1}\left(\frac{10}{12}\right)$$

$$= 56°$$

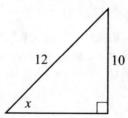

The correct choice is **(4)**.

23. Midpoint E and F form midsegment $\overline{EF}$, which is half the length of the opposite side $\overline{AB}$.

$$EF = \frac{1}{2}AB$$
$$2EF = AB$$
$$2(2x + 8) = 7x - 2$$
$$4x + 16 = 7x - 2$$
$$16 = 3x - 2$$
$$18 = 3x$$
$$x = 6$$

Substitute to find EF and AB.

$$EF = 2(6) + 8$$
$$= 20$$
$$AB = 7(6) - 2$$
$$= 40$$

Because $\triangle ABC$ is equilateral $AB = AC = 40$. AE is half of AC, so AE is 20. FB is also 20. Now add the sides length to find the perimeter of $ABFE$.

$$\text{perimeter} = AE + EF + FB + AB$$
$$= 20 + 20 + 20 + 40$$
$$= 100$$

The correct choice is (**3**).

24. To show a parallelogram is a square, you must show it has a special rectangle property and a special rhombus property. These properties are:

rhombus – adjacent sides congruent, diagonals bisect each other, and diagonals bisect the vertex angles
rectangle – congruent diagonals, and adjacent sides perpendicular.

Perpendicular diagonals is a rectangle property, and congruent adjacent sides is a rhombus property.

The correct choice is (**3**).

PART II

25. A dilation changes side lengths by multiplying them by the scale factor. Since all the side lengths of $\triangle A'BC'$ are half the side lengths of $\triangle ABC$, the triangles cannot be congruent. Congruent figures must have congruent corresponding sides.

26. The area of triangle is found using $A = \frac{1}{2}bh$, where b is the base of the triangle and h is the height. $\triangle PQR$ appears to be a right triangle. We can confirm this by looking at the slopes of QR and PR. If they are negative reciprocals the sides are perpendicular, and the triangle is a right triangle.

$$\text{slope} = \frac{rise}{run}$$

$$\text{slope}_{QR} = -\frac{4}{3}$$

$$\text{slope}_{PR} = \frac{6}{8} = \frac{3}{4}$$

The two slopes are negative reciprocals, making $\triangle PQR$ a right triangle. $\overline{PR}$ is the base of the triangle and $\overline{QR}$ is the height. We can find the lengths using the distance formula.

$$d = \sqrt{\left(x_2 - x_1\right)^2 + \left(y_2 - y_1\right)^2}$$

$$QR = \sqrt{\left(6 - 3\right)^2 + \left(1 - 5\right)^2}$$

$$= \sqrt{\left(3\right)^2 + \left(-4\right)^2}$$

$$= \sqrt{9 + 16}$$

$$= 5$$

$$PR = \sqrt{\left(6 - \left(-2\right)\right)^2 + \left(1 - \left(-5\right)\right)^2}$$

$$= \sqrt{\left(8\right)^2 + \left(6\right)^2}$$

$$= \sqrt{64 + 36}$$

$$= 10$$

Now apply the area formula.

$$A = \frac{1}{2}bh$$

$$= \frac{1}{2}10 \cdot 5$$

$$= 25$$

The area of $\triangle PQR$ is 25.

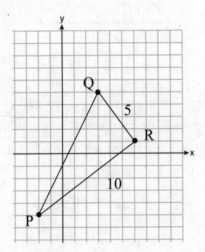

27. The wire, pole and ground form a right triangle as shown. Relative to the 68° angle, the length of 10 along the ground is the adjacent and the unknown wire length is the hypotenuse. Therefore, we should use a cosine ratio to find the wire length.

$$\cos(68°) = \frac{10}{x}$$

$$x\cos(68°) = 10$$

$$x = \frac{10}{\cos(68)}$$

$$= 26.694$$

Rounded to the nearest foot the wire length is 27 ft.

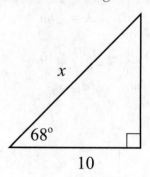

28. The area of a sector is found using $A_{sector} = \pi R^2 \dfrac{\theta}{360}$ where θ is the measure of the central angle in degrees and R is the radius. The central angle is equal in measure to the intercepted arc, which in this case is $72°$. Substitute the $\theta = 72$ and $R = 10$ to find the area.

$$A_{sector} = \pi R^2 \frac{\theta}{360}$$

$$= \pi 10^2 \frac{72}{360}$$

$$= 100\pi$$

The area of the shaded sector is 20π.

29. Start with one pair of coordinates and look for one or more transformations that will map one point to another. Be sure to confirm your transformations work for the other pairs of points. Taking points $A(-2, 5)$ and $T(-5, -2)$, we see that a rotation of $90°$ clockwise about the origin would map A to T since the rotation transforms (x, y) to $(-y, x)$. The same transformation also maps C to U and B to T. A rotation of $90°$ counterclockwise about the origin maps $\triangle ABC$ to $\triangle STU$.

Note that there are numerous other correct answers, such as a rotation of $90°$ counterclockwise about point A followed by a translation 6 down and left 3.

30. Altitude $\overline{PQ}$ drawn to a hypotenuse $\overline{RT}$ forms two triangles similar to the original. From the given side lengths, we are working with $\triangle PRT$ and $\triangle QRP$. Sketch the triangles, matching up corresponding angles. Angles $\angle TPR$ and $\angle PQR$ are congruent right angles. The shared angle at R is also congruent in each. The similarity statement is $\triangle PRT \sim \triangle QRP$. Corresponding sides of similar triangles are proportional, so we can set up a proportion and solve for RQ.

$$\frac{RQ}{PR} = \frac{PR}{RT}$$

$$\frac{RQ}{15} = \frac{15}{17}$$

$$17RQ = 15 \cdot 15$$

$$RQ = \frac{225}{17}$$

$$= 13.23529$$

$RQ = 13.2$ rounded to the nearest tenth.

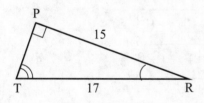
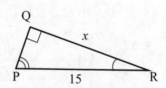

31. Use your compass to measure the radius $\overline{OA}$ by putting the point at O and the pencil at A. Keep the same compass opening and put the point at A and make an arc at B. Move the compass point to B and make an arc at C. Continue making arcs at D, E, and F. If you are accurate, a final arc should bring you back to point A. Connect points A, C and E. $\triangle ACE$ is an equilateral triangle.

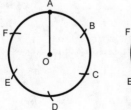
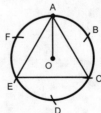

PART III

32. A trapezoid has at least one pair of parallel sides. From the graph, it appears that $\overline{AD}$ and $\overline{BC}$ may be parallel. We can check if they are parallel by calculating the slope of each.

$$\text{slope} = \frac{y_2 - y_1}{x_2 - x_1}$$

$$\text{slope of } \overline{AD} = \frac{0 - 6}{1 - (-1)} = -3$$

$$\text{slope of } \overline{BC} = \frac{-1 - 8}{6 - 3} = -3$$

The slopes of $\overline{AD}$ and $\overline{BC}$ are equal, therefore they are parallel. ABCD is a trapezoid because it has a pair of parallel sides.

We can check if the trapezoid is isosceles by calculating the lengths of the diagonals $\overline{AC}$ and $\overline{BD}$.

$$\text{length} = \sqrt{\left(x_2 - x_1\right)^2 + \left(y_2 - y_1\right)^2}$$

$$\text{length of } \overline{BD} = \sqrt{\left(1 - 3\right)^2 + \left(0 - 8\right)^2}$$

$$= \sqrt{\left(-2\right)^2 + \left(-8\right)^2}$$

$$= \sqrt{4 + 64}$$

$$= \sqrt{68}$$

$$\text{length of } \overline{AC} = \sqrt{\left(6 - (-1)\right)^2 + \left(-1 - 6\right)^2}$$

$$= \sqrt{\left(7\right)^2 + \left(-7\right)^2}$$

$$= \sqrt{49 + 49}$$

$$= \sqrt{98}$$

The lengths of $\overline{BD}$ and $\overline{AC}$ are not equal. The diagonals are not congruent so $ABCD$ is not isosceles.

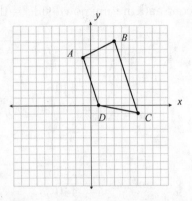

33. The volume of a cylinder is found using $V = \pi R^2 h$ where R is the radius and h is the height. The radius is half the diameter, or 3.25 ft. Since we want our answer in cubic feet, we need to convert the height to feet. 12 inches is equivalent to 1 ft, so the height equals 1 ft.

$$V = \pi R^2 h$$

$$= \pi (3.25)^2 (1)$$

$$= 33.183072 \text{ ft}^3$$

The pool is filled only $\frac{2}{3}$ full, so multiple the volume by $\frac{2}{3}$.

$$V_{water} = \frac{2}{3}(33.183072)$$

$$= 22.122048 \text{ ft}^3$$

The volume of water is 22 ft^3 rounded the nearest cubic foot.

Find the number of gallons by multiplying the volume by the given conversion factor.

$$22 \text{ ft}^3 \cdot 7.48 \, \frac{\text{gallons}}{\text{ft}^3} = 164.56 \text{ gallons}.$$

To the nearest gallon, the amount of water is 165 gallons.

34. The figure can be broken up into the two right triangles shown below. The elevation of the bottom of the blade is y, and the elevation of the top of the blade z. Use two trig ratios to calculate y and z. In each triangle, the 440 ft distance is the adjacent and the unknown side is the opposite, so we use tangent ratios.

calculate y:

$$\tan(30°) = \frac{y}{440}$$
$$y = 440\tan(30°)$$
$$= 254.034118$$

calculate z:

$$\tan(38.8°) = \frac{z}{440}$$
$$z = 440\tan(38.8°)$$
$$= 353.769082$$

The blade length x is found by subtracting y from z.

$$x = z - y$$
$$= 353.769082 - 254.034118$$
$$= 99.734964$$

Rounded to the nearest foot the blade length is 100 ft.

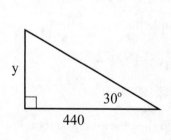
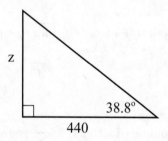

PART IV

35. The strategy for this proof is to show that $\triangle HEA$ is similar to $\triangle TAH$. Corresponding sides in these triangles are proportional, and the proportion leads to $TA \cdot HA = HE \cdot TH$ when cross multiplied. In order to prove the triangles similar, we start by proving $MHTA$ is a parallelogram so we can make use of its parallel sides.

Statements	Reasons
1. $\overline{HM} \cong \overline{AT}, \overline{HT} \cong \overline{AM}$	1. Given
2. $MHTA$ is a parallelogram	2. A quadrilateral with opposite sides congruent is a parallelogram
3. $\overline{HT} \parallel \overline{MA}$	3. Opposite sides of a parallelogram are parallel
4. $\angle THA \cong \angle EAH$	4. Alternate interior angles formed by parallel lines are congruent
5. $\overline{HE} \perp \overline{MEA}$, $\overline{HA} \perp \overline{AT}$	5. Given
6. $\angle HEA$ and $\angle TAH$ are right angles	6. Perpendicular lines form right angles
7. $\angle HEA \cong \angle TAH$	7. Right angles are congruent
8. $\triangle HEA \sim \triangle TAH$	8. AA
9. $\dfrac{TA}{HE} = \dfrac{TH}{HA}$	9. Corresponding sides in similar triangles are proportional
10. $TA \cdot HA = HE \cdot TH$	10. Cross products of a proportion are equal

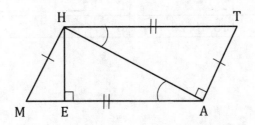

Topic	Question Numbers	Number of Points	Your Points	Your Percentage
1. Basic Angle and Segment Relationships				
2. Angle and Segment Relationships in Triangles and Polygons	21, 23	4		
3. Constructions	31	2		
4. Transformations	1, 4, 5, 29	8		
5. Triangle Congruence	8, 14, 25	6		
6. Lines, Segments, and Circles on the Coordinate Plane	7, 16, 19, 20, 26	10		
7. Similarity	11, 15, 30, 35	12		
8. Trigonometry	9, 22, 27, 34	10		
9. Parallelograms	12, 17, 24	6		
10. Coordinate Geometry Proofs	32	4		
11. Volume and Solids	3, 6, 10, 33	10		
12. Modeling	2	2		
13. Circles	13, 18, 28	6		

Examination
August 2019
Geometry

HIGH SCHOOL MATH REFERENCE SHEET

1 inch = 2.54 centimeters	1 cup = 8 fluid ounces
1 meter = 39.37 inches	1 pint = 2 cups
1 mile = 5280 feet	1 quart = 2 pints
1 mile = 1760 yards	1 gallon = 4 quarts
1 mile = 1.609 kilometers	1 gallon = 3.785 liters
	1 liter = 0.264 gallon
1 kilometer = 0.62 mile	1 liter = 1000 cubic centimeters
1 pound = 16 ounces	
1 pound = 0.454 kilogram	
1 kilogram = 2.2 pounds	
1 ton = 2000 pounds	

Triangle	$A = \frac{1}{2}bh$
Parallelogram	$A = bh$
Circle	$A = \pi r^2$
Circle	$C = \pi d$ or $C = 2\pi r$

General Prisms	$V = Bh$
Cylinder	$V = \pi r^2 h$
Sphere	$V = \frac{4}{3}\pi r^3$
Cone	$V = \frac{1}{3}\pi r^2 h$
Pyramid	$V = \frac{1}{3}Bh$
Pythagorean Theorem	$a^2 + b^2 = c^2$
Quadratic Formula	$x = \dfrac{-b \pm \sqrt{b^2 - 4ac}}{2a}$
Arithmetic Sequence	$a_n = a_1 + (n-1)d$
Geometric Sequence	$a_n = a_1 r^{n-1}$
Geometric Series	$S_n = \dfrac{a_1 - a_1 r^n}{1 - r}$ where $r \neq 1$
Radians	$1 \text{ radian} = \dfrac{180}{\pi} \text{ degrees}$
Degrees	$1 \text{ degree} = \dfrac{\pi}{180} \text{ radians}$
Exponential Growth/Decay	$A = A_0 e^{k(t - t_0)} + B_0$

180 Minutes–35 Questions

PART I

Answer all **24** questions in this part. Each correct answer will receive **2** credits. No partial credit will be allowed. For each statement or question, write in the space provided the numeral preceding the word or expression that best completes the statement or answers the question.   [48 credits]

1 On the set of axes below, triangle *ABC* is graphed. Triangles *A'B'C'* and *A"B"C"*, the images of triangle *ABC*, are graphed after a sequence of rigid motions.

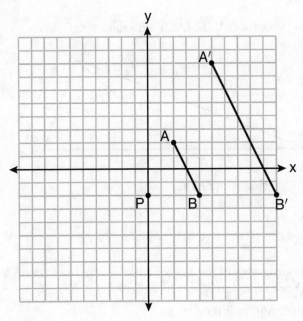

Which statement is always true?

(1) $\overline{PA} \cong \overline{AA'}$

(2) $\overline{AB} \parallel \overline{A'B'}$

(3) $AB = A'B'$

(4) $\frac{5}{2}(A'B') = AB$ 1 _____

2 The coordinates of the vertices of parallelogram *CDEH* are $C(-5,5)$, $D(2,5)$, $E(-1,-1)$, and $H(-8,-1)$. What are the coordinates of *P*, the point of intersection of diagonals $\overline{CE}$ and $\overline{DH}$

(1) $(-2,3)$

(2) $(-2,2)$

(3) $(-3,2)$

(4) $(-3,-2)$ 2 _____

3 The coordinates of the endpoints of $\overline{QS}$ are $Q(-9,8)$ and $S(9,-4)$. Point R is on $\overline{QS}$ such that $QR:RS$ is in the ratio of 1:2. What are the coordinates of point R?

(1) (0,2)   (3) (−3,4)
(2) (3,0)   (4) (−6,6)   3 _____

4 If the altitudes of a triangle meet at one of the triangle's vertices, then the triangle is

(1) a right triangle   (3) an obtuse triangle
(2) an acute triangle   (4) an equilateral triangle   4 _____

5 In the diagram below of $\triangle ACD$. $\overline{DB}$ is a median to $\overline{AC}$, and $\overline{AB} \simeq \overline{DB}$.

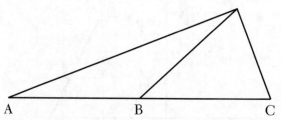

If m∠DAB = 32°, what is m∠BDC?

(1) 32°   (3) 58°
(2) 52°   (4) 64°   5 _____

6 What are the coordinates of the center and the length of the radius of the circle whose equation is $x^2 + y^2 = 8x - 6y + 39$?

(1) center (−4,3) and radius 64
(2) center (4,−3) and radius 64
(3) center (−4,3) and radius 8
(4) center (4,−3) and radius 8   6 _____

7 In the diagram below of parallelogram $ABCD$, $\overline{AFGB}, \overline{CF}$ bisects $\angle DCB$, $\overline{DG}$ bisects $\angle ADC$, and $\overline{CF}$ and $\overline{DG}$ intersect at E.

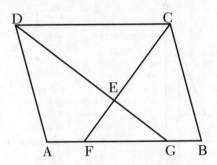

If $m\angle B = 75°$, then the measure of $\angle EFA$ is

(1) 142.5°

(3) 52.5°

(2) 127.5°

(4) 37.5°   7 _____

8 What is an equation of a line that is perpendicular to the line whose equation is $2y + 3x = 1$?

(1) $y = \frac{2}{3}x + \frac{5}{2}$

(3) $y = -\frac{2}{3}x + 1$

(2) $y = \frac{3}{2}x + 2$

(4) $y = -\frac{3}{2}x + \frac{1}{2}$ 8 _____

9 Triangles *ABC* and *RST* are graphed on the set of axes below.

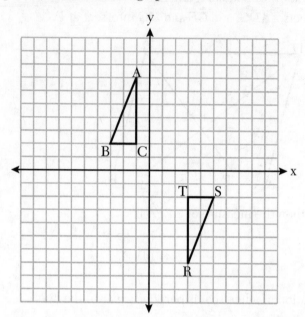

Which sequence of rigid motions will prove △*ABC* ≅ △*RST*?

(1) a line reflection over $y = x$

(2) a rotation of 180° centered at (1,0)

(3) a line reflection over the *x*-axis followed by a translation of
6 units right

(4) a line reflection over the *x*-axis followed by a line reflection
over $y = x$ 9 ____

10 If the line represented by $y = -\frac{1}{4}x - 2$ is dilated by a scale factor
of 4 centered at the origin, which statement about the image is
true?

(1) The slope is $-\frac{1}{4}$ and the *y*-intercept is −8.

(2) The slope is $-\frac{1}{4}$ and the *y*-intercept is −2.

(3) The slope is −1 and the *y*-intercept is −8.

(4) The slope is −1 and the *y*-intercept is −2.   10 ____

11 Square *MATH* has a side length of 7 inches. Which three-dimensional object will be formed by continuously rotating square *MATH* around side $\overline{AT}$?

   (1) a right cone with a base diameter of 7 inches

   (2) a right cylinder with a diameter of 7 inches

   (3) a right cone with a base radius of 7 inches

   (4) a right cylinder with a radius of 7 inches   11 _____

12 Circle *O* with a radius of 9 is drawn below. The measure of central angle *AOC* is 120°.

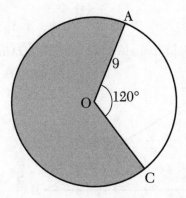

What is the area of the shaded sector of circle *O*?

   (1) 6π (3) 27π

   (2) 12π (4) 54π 12 _____

13 In quadrilateral *QRST*, diagonals $\overline{QS}$ and $\overline{RT}$ intersect at *M*. Which statement would always prove quadrilateral *QRST* is a parallelogram?

   (1) $\angle TQR$ and $\angle QRS$ are supplementary.

   (2) $\overline{QM} \cong \overline{SM}$ and $\overline{QT} \cong \overline{RS}$

   (3) $\overline{QR} \cong \overline{TS}$ and $\overline{QT} \cong \overline{RS}$

   (4) $\overline{QR} \cong \overline{TS}$ and $\overline{QT} \parallel \overline{RS}$ 13 _____

14 A standard-size golf ball has a diameter of 1.680 inches. The material used to make the golf ball weighs 0.6523 ounce per cubic inch. What is the weight, to the *nearest hundredth of an ounce*, of one golf ball?

(1) 1.10   (3) 2.48
(2) 1.62   (4) 3.81   14 ____

15 Chelsea is sitting 8 feet from the foot of a tree. From where she is sitting, the angle of elevation of her line of sight to the top of the tree is 36°. If her line of sight starts 1.5 feet above ground, how tall is the tree, to the *nearest foot*?

(1) 8   (3) 6
(2) 7   (4) 4   15 ____

16 In the diagram below of right triangle ABC, altitude $\overline{CD}$ intersects hypotenuse $\overline{AB}$ at D.

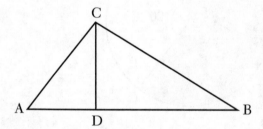

Which equation is always true?

(1) $\dfrac{AD}{AC} = \dfrac{CD}{BC}$ (3) $\dfrac{AC}{CD} = \dfrac{BC}{CD}$

(2) $\dfrac{AD}{CD} = \dfrac{BD}{CD}$ (4) $\dfrac{AD}{AC} = \dfrac{AC}{BD}$ 16 ____

17 A countertop for a kitchen is modeled with the dimensions shown below. An 18-inch by 21-inch rectangle will be removed for the installation of the sink.

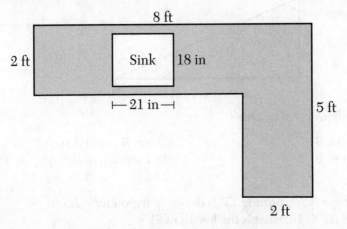

What is the area of the top of the installed countertop, to the *nearest square foot*?

(1) 26   (3) 22

(2) 23   (4) 19   17 _____

18 In the diagram below, $\overline{BC}$ connects points B and C on the congruent sides of isosceles triangle ADE, such that $\triangle ABC$ is isosceles with vertex angle A.

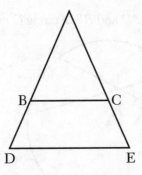

If $AB = 10$, $BD = 5$, and $DE = 12$, what is the length of $\overline{BC}$?

(1) 6   (3) 8

(2) 7   (4) 9   18 _____

19 In △ABC below, angle C is a right angle.

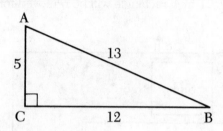

Which statement must be true?

(1) sin A = cos B   (3) sin B = tan A
(2) sin A = tan B   (4) sin B = cos B   19 _____

20 In right triangle RST, altitude $\overline{TV}$ is drawn to hypotenuse $\overline{RS}$. If RV = 12 and RT = 18, what is the length of $\overline{SV}$?

(1) 6√5   (3) 6√6
(2) 15   (4) 27   20 _____

21 What is the volume, in cubic centimeters, of a right square pyramid with base edges that are 64 cm long and a slant height of 40 cm?

(1) 8192.0   (3) 32,768.0
(2) 13,653.$\overline{3}$ (4) 54,613.$\overline{3}$ 21 _____

22 In the diagram below, chords $\overline{PQ}$ and $\overline{RS}$ of circle O intersect at T.

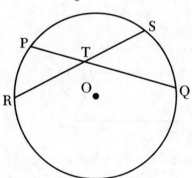

Which relationship must always be true?

(1) RT = TQ   (3) RT + TS = PT + TQ
(2) RT = TS   (4) RT × TS = PT × TQ   22 _____

23 A rhombus is graphed on the set of axes below.

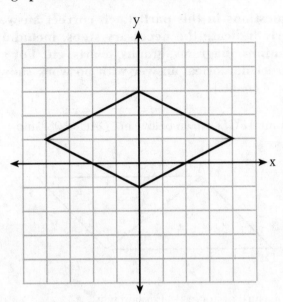

Which transformation would carry the rhombus onto itself?

(1) 180° rotation counterclockwise about the origin

(2) reflection over the line $y = \frac{1}{2}x + 1$

(3) reflection over the line $y = 0$

(4) reflection over the line $x = 0$ 23 _____

24 A 15-foot ladder leans against a wall and makes an angle of 65°
with the ground. What is the horizontal distance from the wall to
the base of the ladder, to the *nearest tenth of a foot*?

(1) 6.3   (3) 12.9

(2) 7.0   (4) 13.6   24 _____

PART II

Answer all 7 questions in this part. Each correct answer will receive 2 credits. Clearly indicate the necessary steps, including appropriate formula substitutions, diagrams, graphs, charts, etc. For all questions in this part, a correct numerical answer with no work shown will receive only 1 credit. [14 credits]

25 In parallelogram *ABCD* shown below, m∠*DAC* = 98° and m∠*ACD* = 36°.

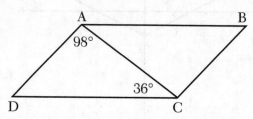

What is the measure of angle B? Explain why.

26 An airplane took off at a constant angle of elevation. After the plane traveled for 25 miles, it reached an altitude of 5 miles, as modeled below.

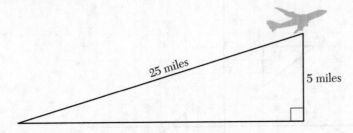

To the *nearest tenth of a degree*, what was the angle of elevation?

27 On the set of axes below, $\triangle ABC \cong \triangle DEF$.

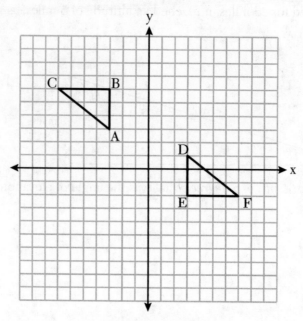

Describe a sequence of rigid motions that maps $\triangle ABC$ onto $\triangle DEF$.

28 The vertices of △ABC have coordinates $A(-2,-1)$, $B(10,-1)$, and $C(4,4)$. Determine and state the area of △ABC. [The use of the set of axes below is optional.]

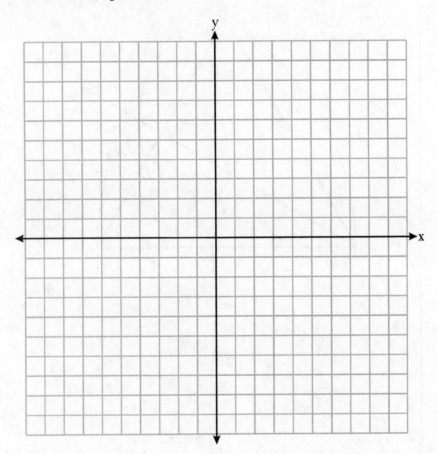

29 Using the construction below, state the degree measure of ∠CAD.
    Explain why.

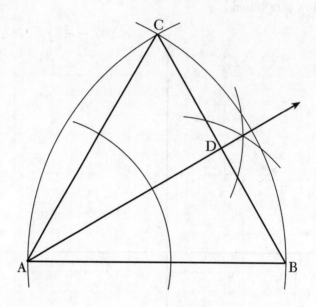

30 In the diagram below of circle K, secant $\overline{PLKE}$ and tangent $\overline{PZ}$ are drawn from external point P.

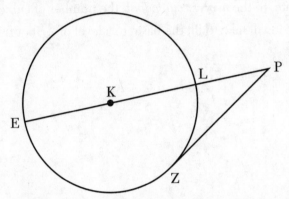

If $m\widehat{LZ} = 56°$, determine and state the degree measure of angle P.

31 A large water basin is in the shape of a right cylinder. The inside of the basin has a diameter of $8\frac{1}{4}$ feet and a height of 3 feet. Determine and state, to the *nearest cubic foot*, the number of cubic feet of water that it will take to fill the basin to a level of $\frac{1}{2}$ foot from the top.

PART III

Answer all 3 questions in this part. Each correct answer will receive 4 credits. Clearly indicate the necessary steps, including appropriate formula substitutions, diagrams, graphs, charts, etc. For all questions in this part, a correct numerical answer with no work shown will receive only 1 credit.   [12 credits]

32 Triangle *ABC* is shown below. Using a compass and straightedge, construct the dilation of △*ABC* centered at *B* with a scale factor of 2. [Leave all construction marks.]

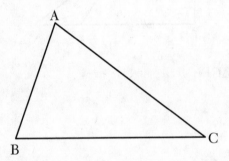

Is the image of △*ABC* similar to the original triangle? Explain why.

33 In the diagram below, $\triangle ABE \cong \triangle CBD$.

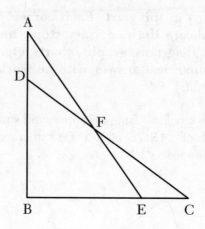

Prove: $\triangle AFD \cong \triangle CFE$

34 A cargo trailer, pictured below, can be modeled by a rectangular prism and a triangular prism. Inside the trailer, the rectangular prism measures 6 feet wide and 10 feet long. The walls that form the triangular prism each measure 4 feet wide inside the trailer. The diagram below is of the floor, showing the inside measurements of the trailer.

Cargo Trailer

Cargo Trailer Floor

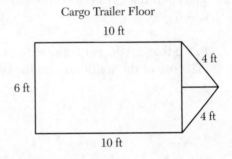

If the inside height of the trailer is 6.5 feet, what is the total volume of the inside of the trailer, to the *nearest cubic foot*?

PART IV

Answer the question in this part. A correct answer will receive 6 credits. Clearly indicate the necessary steps, including appropriate formula substitutions, diagrams, graphs, charts, etc. A correct numerical answer with no work shown will receive only 1 credit.   [6 credits]

35 The coordinates of the vertices of $\triangle ABC$ are $A(1,2)$, $B(-5,3)$, and $C(-6,-3)$.

Prove that $\triangle ABC$ is isosceles.
[The use of the set of axes on the next page is optional.]

State the coordinates of point D such that quadrilateral $ABCD$ is a square.

Question 35 is continued on the next page.

Question 35 continued

Prove that your quadrilateral *ABCD* is a square.
[The use of the set of axes below is optional.]

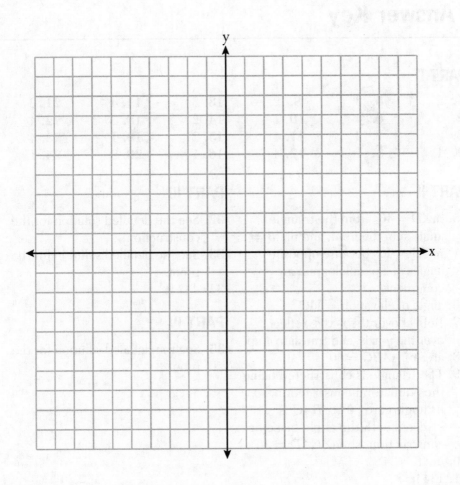

Answers
August 2019
Geometry

Answer Key

PART I

1. 2	**5.** 3	**9.** 2	**13.** 3	**17.** 4	**21.** 3
2. 3	**6.** 4	**10.** 1	**14.** 2	**18.** 3	**22.** 4
3. 3	**7.** 2	**11.** 4	**15.** 2	**19.** 1	**23.** 4
4. 1	**8.** 1	**12.** 4	**16.** 1	**20.** 2	**24.** 1

PART II

25. m$\angle D = 46°$ from the triangle angle sum theorem, and m$\angle B =$ m$\angle D = 46°$ because opposite angles of a parallelogram are congruent.

26. angle of elevation $= 11.5°$

27. Reflect over the x-axis, reflect over the y-axis, and translate 4 up.

28. area of $\triangle ABC = 30$

29. The triangle is constructed using the equilateral construction, so its angles are 60°. $\angle BAC$ is bisected with the angle bisector construction, so m$\angle CAD = 30°$.

30. m$\angle P = 34°$

31. 134 ft^3

PART III

32. See the detailed solution for the construction.

33. See the detailed solution for the proof.

34. 442 ft^3

PART IV

35. See the detailed solution for the proof.

In **PARTS II–IV**, you are required to show how you arrived at your answers. For sample methods of solutions, see the *Answers Explained* section.

Answers Explained

PART I

1. The dilation of a line segment is always parallel to the original segment, so $\overline{AB} \parallel \overline{A'B'}$. Examining the other choices, we see that choice (1) is incorrect because $\overline{PA} \cong \overline{AA'}$ implies a 2:1 ratio instead of the 5:2 ratio. Choices (3) and (4) are incorrect because. $A'B' = \frac{5}{2} AB$.

   The correct choice is **(2)**.

2. The diagonals of a parallelogram bisect each other, so the point of intersection must be the midpoint of the diagonals. Using the coordinates $(-5, 5)$ and $(-1, -1)$ for diagonal $\overline{CE}$, apply the midpoint formula.

$$\text{midpoint} = \left(\frac{x_1 + x_2}{2}, \frac{y_1 + y_2}{2} \right)$$

$$= \left(\frac{-5 + (-1)}{2}, \frac{5 + (-1)}{2} \right)$$

$$= (-3, 2)$$

   The correct choice is **(3)**. Note that you would get the same result using diagonal $\overline{DH}$.

3. The coordinates of the point that divide a segment in a specified ratio can be calculated using the following formula:

$$\text{ratio} = \frac{x - x_1}{x_2 - x} \qquad \text{ratio} = \frac{y - y_1}{y_2 - y}$$

   where the coordinates of $Q(-9, 8)$ are (x_1, y_1) the coordinates of $S(9, -4)$ are (x_2, y_2).

$$\frac{1}{2} = \frac{x - (-9)}{9 - x} \qquad\qquad \frac{1}{2} = \frac{y - 8}{-4 - y}$$

$$2x + 18 = 9 - x \qquad\qquad 2y - 16 = -4 - y$$

$$3x = -9 \qquad\qquad\qquad 3y = 12$$

$$x = -3 \qquad\qquad\qquad y = 4$$

   The coordinates of R are $(-3, 4)$.

   The correct choice is **(3)**.

4. Altitudes of a triangle always form a right angle with the opposite side. If the altitudes meet at a vertex, then the triangle must be a right triangle.

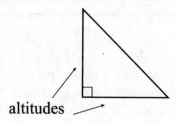

The correct choice is (**1**).

5. Since $\overline{AB} \cong \overline{BD}$, we know $\triangle ABD$ is isosceles and $\angle DAB \cong \angle ADB$. Apply the angle sum theorem in $\triangle ABD$ to find m$\angle ABD$.

$$m\angle ABD + m\angle DAB + m\angle ADB = 180°$$
$$m\angle ABD + 32° + 32° = 180°$$
$$m\angle ABD = 116°.$$

Next, we see that $\angle ABD$ and $\angle CBD$ are a linear pair and sum to 180°.

$$m\angle ABD + m\angle CBD = 180°$$
$$116° + m\angle CBD = 180°$$
$$m\angle CBD = 64°.$$

A median divides a side of a triangle into two congruent segments, so $\overline{AB} \cong \overline{BC}$. Combining this with $\overline{AB} \cong \overline{BD}$, we can conclude that $\overline{BC} \cong \overline{BD}$. We now know that $\triangle BDC$ is isosceles with $\angle BDC \cong \angle BCD$. Apply the triangle sum theorem to find m$\angle BDC$.

$$m\angle BDC + m\angle BCD + m\angle CBD = 180°$$
$$2\, m\angle BDC + 64° = 180°$$
$$2\, m\angle BDC = 116°$$
$$m\angle BDC = 58°$$

The correct choice is (**3**).

6. Use the completing the square procedure to convert the equation to center-radius form, $(x - h)^2 + (y - k^2) = R^2$. The center has coordinates (h, k) and the radius is R. Start by rearranging the equation so all the x and y terms are grouped together on the left.

$$x^2 + y^2 = 8x - 6y + 39$$

$$x^2 - 8x + y^2 + 6y = 39$$

Now find the constant terms we need to complete the square:

constant for the x-terms $= (\frac{1}{2} \cdot \text{coefficient of } x)^2$

$$= \left(\frac{1}{2} \cdot (-8)\right)^2$$

$$= 16$$

constant for the y-terms $= (\frac{1}{2} \cdot \text{coefficient of } y)^2$

$$= \left(\frac{1}{2} \cdot 6\right)^2$$

$$= 9$$

Add the required constant to each side of the equation and factor.

$$x^2 - 8x + 16 + y^2 + 6y + 9 = 39 + 16 + 9$$

$$(x - 4)^2 + (y + 3)^2 = 64$$

The values of h and k are 4 and 3. Be careful with the signs because h and k are the values subtracted from x and y. The center is located at $(4, -3)$. Since $R^2 = 64$, the radius is equal to $\sqrt{64}$, or 8.

The correct choice is **(4)**.

7. Consecutive angles of a parallelogram are supplementary, so we can find $m\angle BCD$.

$$m\angle BCD + m\angle B = 180°$$

$$m\angle BCD + 75° = 180°$$

$$m\angle BCD = 105°$$

Next, use angle bisector $\overline{CF}$ to find m$\angle DCF$.

$$m\angle DCF = \frac{1}{2} \, m\angle BCD$$
$$= \frac{1}{2}(105°)$$
$$= 52.5°$$

$\angle DCF$ and $\angle CFB$ are congruent alternate interior angles, so m$\angle CFG$ is also equal to 52.5°. Finally, $\angle EFA$ and $\angle CFG$ are a linear pair that sum to 180°.

$$m\angle EFA + m\angle CFG = 180°$$
$$m\angle EFA + 52.5° = 180°$$
$$m\angle EFA = 127.5°$$

The correct choice is (**2**).

8. First, identify the slope by rewriting the equation in $y = mx + b$ form.

$$2y + 3x = 1$$
$$2y = -3x + 1$$
$$y = -\frac{3}{2}x + \frac{1}{2}$$

The slope of the given line is $-\frac{3}{2}$. Slopes of perpendicular lines are negative reciprocals, which we can find by changing the sign and flipping the fraction. The negative reciprocal of $-\frac{3}{2}$ is $\frac{2}{3}$. The only choice with slope of $\frac{2}{3}$ is $y = \frac{2}{3}x + \frac{5}{2}$.

The correct choice is (**1**).

9.

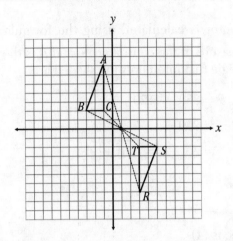

Two figures are congruent if a sequence of rigid motions maps one figure onto another. To check if a 180° rotation was used, sketch segments from each vertex of △ABC to the corresponding vertex of △RST as shown in the figure. All three segments are concurrent at (1, 0), so this point is the center of a 180° rotation that maps △ABC to △RST.

The correct choice is **(2)**.

10. When a line is dilated about the origin, the slope remains the same, but the y-intercept is multiplied by the scale factor. The slope remains $-\frac{1}{4}$. The y-intercept transforms to $4 \cdot (-2)$, or -8.

The correct choice is **(1)**.

11. A square rotated about any of its sides always sweeps out a cylinder. The side length of the square is the height, as well as the radius of the cylinder. The solid is a cylinder with a radius of 7 inches.

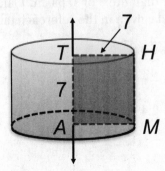

The correct choice is **(4)**.

12. The area of a sector is calculated using the formula $A_{\text{sector}} = \pi R^2 \dfrac{\theta}{360}$, where θ is the measure of the central angle and R is the radius. First, calculate the central angle that intercepts the shaded arc.

$$\theta = 360° - 120°$$
$$= 240°$$
$$= A_{\text{sector}} = \pi R^2 \frac{\theta}{360}$$
$$= \pi \cdot 9^2 \cdot \frac{240}{360}$$
$$= 54\pi$$

The correct choice is (**4**).

13. One possible property that would prove a quadrilateral is a parallelogram is congruent opposite sides. For quadrilateral $QRST$, congruent opposite sides would be $\overline{QR} \cong \overline{TS}$ and $\overline{QT} \cong \overline{RS}$.

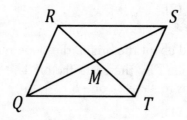

The correct choice is (**3**).

14. The golf ball can be modeled as a sphere with a diameter of 1.680 in. Its radius is one half the diameter, or 0.84 in. Calculate the volume of the sphere using the formula given in the reference table.

$$V = \frac{4}{3}\pi R^3$$
$$= \frac{4}{3}\pi (0.84)^3$$
$$= 2.482713 \text{ in}^3$$

The weight of an object is equal to the product of its volume and density.

$$\text{weight} = \text{volume} \cdot \text{density}$$

$$= 2.482713 \text{ in}^3 \left(0.6523 \frac{\text{ounces}}{\text{in}^3} \right)$$

$$= 1.62 \text{ ounces}$$

The correct choice is **(2)**.

15. The tree and Chelsea's line of sight form a right triangle as shown in the figure. Relative to the 36° angle of elevation, the 8-foot distance is the adjacent and the observed tree height is the opposite. A tangent ratio can be used to find the unknown height.

$$\tan(36°) = \frac{\text{opposite}}{\text{adjacent}}$$

$$\tan(36°) = \frac{x}{8}$$

$$x = 8 \tan(36°)$$

$$= 5.8123 \text{ ft}$$

Add the 1.5-ft distance from the ground to Chelsea's line of sight to get the total height.

$$\text{total height} = 5.8123 + 1.5$$

$$= 7.3123 \text{ ft}$$

To the nearest foot the height of the tree is 7 feet.

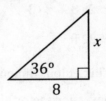

The correct choice is **(2)**.

16. When an altitude is drawn to the hypotenuse of a right triangle, three similar triangles are formed. Sketch the three triangles separately and label the congruent angles. We know that corresponding sides of similar triangles are congruent, so we can write the following proportion:

$$\frac{AD}{AC} = \frac{CD}{BC}$$

$\overline{AD}$ and $\overline{CD}$ are corresponding sides of the small and medium triangles. $\overline{AC}$ and $\overline{BC}$ are also corresponding sides of the same two triangles. None of the other choices match up corresponding sides of the same two triangles.

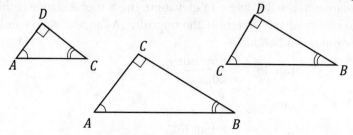

The correct choice is **(1)**.

17. The countertop can be broken up into three rectangles—a solid 2 ft by 8 ft rectangle, a solid 2 ft by 3 ft rectangle, and a 21 in × 18 in cutout. Calculate the area of two solid rectangles.

$$\text{Area}_{\text{solid}} = \text{length}_1 \cdot \text{width}_1 + \text{length}_2 \cdot \text{width}_2$$

$$= 2 \cdot 8 + 2 \cdot 3$$

$$= 22 \text{ ft}^2$$

Before finding its area, the sink cutout dimensions need to be converted to feet by multiplying by the conversion factor $\dfrac{1 \text{ ft}}{12 \text{ in}}$.

$$\text{Sink width} = 21 \text{ in} \cdot \frac{1 \text{ ft}}{12 \text{ in}} \qquad\qquad \text{Sink length} = 18 \text{ in} \cdot \frac{1 \text{ ft}}{12 \text{ in}}$$

$$= 1.75 \text{ ft} \qquad\qquad\qquad\qquad = 1.5 \text{ ft}$$

$$\text{Area}_{\text{sink}} = (1.75)(1.5)$$

$$= 2.625 \text{ ft}^2$$

Now subtract the sink cutout for the sink from the solid area.

$$\text{Area}_{\text{countertop}} = 22 - 2.625$$

$$= 19.375 \text{ ft}^2$$

Rounded to the nearest square foot the area is 19 ft^2.

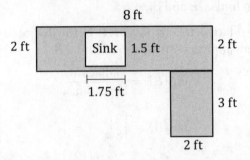

The correct choice is (**4**).

18. Using the definition of isosceles triangles, we know $\overline{AB} \cong \overline{AC}$ and $\overline{AD} \cong \overline{AE}$. Dividing these congruence statements, we get $\frac{AB}{AD} \cong \frac{AC}{AE}$. The shared angle $\angle A$ is congruent in each triangle as well, so $\triangle BAC$ is similar to $\triangle DAE$ by the SAS postulate. Apply the theorem that corresponding sides in similar triangles are proportional.

$$\frac{BA}{DA} = \frac{BC}{DE}$$

$$\frac{10}{10 + 5} = \frac{BC}{12}$$

$$\frac{10}{15} = \frac{BC}{12}$$

$$15BC = 120$$

$$BC = 8$$

The correct choice is (**3**).

19. Given two complementary angles A and B, the co-function relationship states that $\sin(A) = \cos(B)$. In right triangle ABC, angle C is a right triangle, so $\angle A$ and $\angle B$ are complementary. Therefore, $\sin(A) = \cos(B)$.

The correct choice is (**1**).

20. There are several ways to approach this problem. One method is to apply trigonometry to find $\angle R$, and then RS.

Using $\triangle TVR$, relative to $\angle R$ side $\overline{RT}$ is the hypotenuse and $\overline{RV}$ is the adjacent. We can apply a cosine ratio to find $\angle R$.

$$\cos(R) = \frac{\text{adjacent}}{\text{hypotenuse}}$$

$$\cos(R) = \frac{12}{18}$$

$$R = \cos^{-1}\left(\frac{12}{18}\right)$$

$$R = 48.189685°$$

Now use $\triangle TRS$ to find RS. Relative to $\angle R$ side $\overline{RS}$ is the hypotenuse and side $\overline{TR}$ is the adjacent. Use a cosine ratio to find $\overline{RS}$.

$$\cos(R) = \frac{\text{adjacent}}{\text{hypotenuse}}$$

$$\cos(48.189685) = \frac{18}{RS}$$

$$RS = \frac{18}{\cos(48.189685)}$$

$$= 27$$

From the figure,

$$SV = RS - 12$$

$$= 27 - 12$$

$$= 15$$

The correct choice is (**2**).

21. The height, h, of the pyramid forms a right triangle with the base and the 40 cm slant height. Use the Pythagorean theorem to find the height h. Be sure to use half of the side length for the bottom of the right triangle.

$$a^2 + b^2 = c^2$$
$$h^2 + 32^2 = 40^2$$
$$h^2 + 1024 = 1600$$
$$h^2 = 576$$
$$h = 24$$

Now find the volume of the pyramid using the volume formula from the reference table. The area of the base, B, is the area of the square base.

$$V = \frac{1}{3}Bh$$
$$= \frac{1}{3}\left(64^2\right)24$$
$$= 32{,}768$$

The volume of the pyramid is $32{,}768$ cm^3.

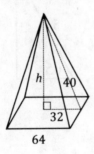

The correct choice is (**3**).

22. When two chords intersect within a circle, the products of the parts of the chords are equal. In this case $RT \cdot TS = PT \cdot TQ$.

The correct choice is (**4**).

23. The rhombus is symmetric about the y-axis, so a reflection over that axis will map the rhombus onto itself. The equation of the y-axis is $x = 0$.

The correct choice is (**4**).

24. The ladder, wall, and ground form a right triangle. We can use trigonometry to calculate the distance from the ladder to the wall. Relative to the 65° angle, the ladder is the hypotenuse and the distance to the wall is the adjacent. This indicates a cosine ratio.

$$\cos(65°) = \frac{\text{adjacent}}{\text{hypotenuse}}$$

$$\cos(65°) = \frac{x}{15}$$

$$x = 15\cos(65°)$$

$$= 6.33927$$

Rounded to the nearest foot, the distance to the wall is 6.3 ft.

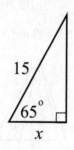

The correct choice is (**1**).

PART II

25. Two of the three angles are given in △ADC. Apply the angle sum theorem to find ∠D.

$$m\angle D + 36° + 98° = 180°$$

$$m\angle D + 134° = 180°$$

$$m\angle D = 46°$$

Opposite angles of a parallelogram are congruent, so m∠B = m∠D. Therefore, m∠B = 46°.

26. The angle of elevation is the angle formed between the ground and the 25-mile flight path. Relative to the angle of elevation, the 25-mile flight path is the hypotenuse and the 5-mile altitude is the opposite. This indicates using a sine ratio. Set up the correct ratio and then apply the inverse sine function to find the angle.

$$\sin(\theta) = \frac{5}{25}$$

$$\theta = \sin^{-1}\left(\frac{5}{25}\right)$$

$$= 11.53695°$$

Rounded to the nearest tenth of a degree, the angle of elevation is 11.5°.

27. There are many possible sequences of transformations that will map △ABC to △DEF. One possible sequence is two reflections followed by a translation. First, reflect △ABC over the x-axis to triangle I. Then, reflect triangle I over the y-axis to triangle II. Finally, translate triangle II up 4 to △DEF.

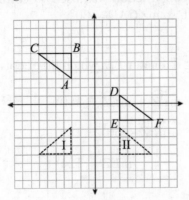

28. After graphing the triangle, sketch altitude $\overline{CD}$ to side $\overline{AB}$. We can calculate their lengths by counting grid units since they are vertical and horizontal segments. We find that $AB = 12$ and $CD = 5$. Now apply the formula for the area of a triangle.

$$A = \frac{1}{2}bh$$

$$= \frac{1}{2}(12)(5)$$

$$= 30$$

The area of $\triangle ABC$ is 30.

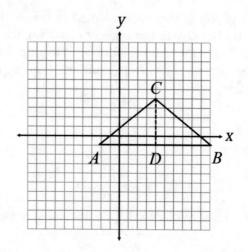

29. $\triangle ABC$ is constructed using the equilateral triangle construction. Since all angles of an equilateral triangle measure 60°, we know that $\mathrm{m}\angle CAB = 60°$. The angle bisector construction was used to construct $\overrightarrow{AD}$. Bisector $\overrightarrow{AD}$ divides $\angle CAB$ in half, so $\mathrm{m}\angle CAD = 30°$.

30. First find $\mathrm{m}\overset{\frown}{EZ}$ using diameter $\overline{LE}$, which intercepts a 180° semicircle.

$$\mathrm{m}\overset{\frown}{EZ} + \mathrm{m}\overset{\frown}{LZ} = 180°$$

$$\mathrm{m}\overset{\frown}{EZ} + 56° = 180°$$

$$\mathrm{m}\overset{\frown}{EZ} = 124°$$

When a tangent and secant intersect, the angle formed is equal to one-half the difference of the intercepted arcs.

$$m\angle P = \frac{1}{2}\left(m\widehat{EZ} - m\widehat{LZ}\right)$$

$$= \frac{1}{2}(124° - 56°)$$

$$= 34°$$

31. Calculate the height of the water in the cylinder by subtracting $\frac{1}{2}$ ft from the height of the basin.

$$h_{water} = 3 - \frac{1}{2}$$

$$= 2.5 \text{ ft}$$

The radius of the cylinder is half the diameter, or 4.125 ft.

Next, calculate the volume of the water using the volume formula for a cylinder.

$$V = \pi R^2 h$$

$$= \pi(4.125^2)(2.5)$$

$$= 133.64041 \text{ ft}^3$$

To the nearest cubic foot, the volume of water is 134 ft^3.

PART III

32. (a) With the compass point at B, open the compass so the pencil is at A and make an arc to show that you measured that distance.

(b) Move the point to A and, with the same opening, make a second arc.

(c) Extend $\overline{BA}$ to the second arc. Label the intersection point E.

(d) Extend $\overline{BC}$ to F by repeating steps (a)–(c).

$\triangle EBF$ is a dilation of $\triangle ABC$ by a scale factor of 2 centered at B. The image is similar to the original because dilations are a similarity transformation. They increase length without changing any of the angle measures.

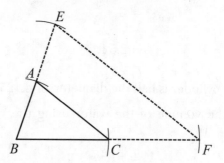

33. Prove the two triangles congruent by AAS. We will use a pair of congruent angles from the given congruent triangles, a pair of vertical angles, and sides $\overline{AD}$ and $\overline{CE}$. The sides are shown to be congruent by subtracting two pairs of congruent sides from the given triangles.

Statement	Reason
1. $\triangle ABE \cong \triangle CBD$	1. Given
2. $\angle A \cong \angle C$	2. CPCTC
3. $\angle AFD \cong \angle CFE$	3. Vertical angles are congruent
4. $\overline{AB} \cong \overline{CB}$	4. CPCTC
5. $\overline{BD} \cong \overline{BE}$	5. CPCTC
6. $\overline{AB} - \overline{BD} \cong \overline{CB} - \overline{BE}$	6. Subtraction Postulate
7. $\overline{AD} \cong \overline{CE}$	7. Partition Postulate
8. $\triangle AFD \cong \triangle CFE$	8. AAS

34. The volume of a prism is calculated using $V = Bh$, where B is the area of the base and h is the height of the prism. First, calculate the volume of the rectangular prism. The base is a 10 ft by 6 ft rectangle and the height is 6.5 ft.

$$B = 10 \cdot 6$$
$$= 60 \text{ ft}^2$$

$$V = Bh$$
$$= (60)(6.5)$$
$$= 390 \text{ ft}^3$$

The base of the triangular prism is an isosceles triangle with 4-ft legs and a base of 6 ft. To find its area we need to calculate the altitude, a, of the triangle. The altitude forms two right triangles as shown in the figure, and we can use the Pythagorean theorem to calculate a.

$$a^2 + b^2 = c^2$$
$$a^2 + 3^2 = 4^2$$
$$a^2 = 7$$
$$a = \sqrt{7}$$

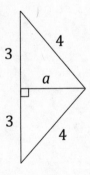

The area of the triangle base can now be calculated.

$$B = \frac{1}{2}(6)\left(\sqrt{7}\right)$$
$$= 3\sqrt{7}$$

Now calculate the volume of the triangular prism.

$$V = Bh$$
$$= \left(3\sqrt{7}\right)(6.5)$$

$$= 51.592 \text{ ft}^2$$

Finally combine the two volumes.

$$V_{\text{total}} = 390 \text{ ft}^3 + 51.592 \text{ ft}^3$$

$$441.592 \text{ ft}^3$$

Rounded to the nearest cubic foot the volume is 442 ft^3.

PART IV

35. An isosceles triangle has two congruent sides, so the proof requires us to show the lengths of two sides are equal. $\overline{AB}$ and $\overline{BC}$ appear to be the congruent sides, so start with those.

Length of AB:

$$d = \sqrt{\left(x_1 - x_2\right)^2 + \left(y_1 - y_2\right)^2}$$
$$= \sqrt{\left(1 - (-5)\right)^2 + (2 - 3)^2}$$
$$= \sqrt{37}$$

Length of BC:

$$d = \sqrt{\left(x_1 - x_2\right)^2 + \left(y_1 - y_2\right)^2}$$
$$= \sqrt{\left(-5 - (-6)\right)^2 + (3 - (-3))^2}$$
$$= \sqrt{37}$$

$\overline{AB}$ and $\overline{BC}$ have the same length, so they are congruent and $\triangle ABC$ is isosceles.

To find point D that completes the square, count the horizontal and vertical translations from point B to point A. Point A is 6 units to the right and 1 unit down from point B. Count those same distances from point C to locate point D. Starting from C $(-6, -3)$, 6 units right and 1 unit down puts point D at $(0, -4)$.

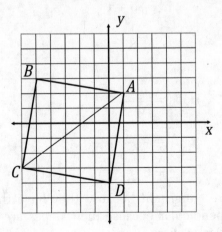

We need to demonstrate one parallelogram property, one rhombus property, and one rectangle property. One possible approach is to show:

- opposite sides are congruent (parallelogram)
- consecutive sides are congruent (rhombus)
- consecutive sides form a right angle (rectangle)

Since we already have the lengths of $\overline{AB}$ and $\overline{BC}$, we just need $\overline{CD}$ and $\overline{AD}$.

Length of CD:

$$d = \sqrt{\left(x_1 - x_2\right)^2 + \left(y_1 - y_2\right)^2}$$
$$= \sqrt{\left(-6 - 0\right)^2 + \left(-3 - (-4)\right)^2}$$
$$= \sqrt{37}$$

Length of AD:

$$d = \sqrt{\left(x_1 - x_2\right)^2 + \left(y_1 - y_2\right)^2}$$
$$= \sqrt{\left(1 - 0\right)^2 + \left(2 - (-4)\right)^2}$$
$$= \sqrt{37}$$

Now calculate the slopes of $\overline{AB}$ and $\overline{BC}$.

Slope of AB:

$$\text{slope} = \frac{y_2 - y_1}{x_2 - x_1}$$
$$= \frac{3 - 2}{-5 - 1}$$
$$= -\frac{1}{6}$$

Slope of BC:

$$\text{slope} = \frac{y_2 - y_1}{x_2 - x_1}$$
$$= \frac{-3 - 3}{-6 - (-5)}$$
$$= 6$$

All four sides are congruent because they have the same lengths. $\overline{AB}$ and $\overline{BC}$ are perpendicular because their slopes are negative reciprocals. Therefore, $ABCD$ is a square.

The Geometry Learning Standards

The New York State Geometry Regents is aligned to the Geometry Learning Standards. All topics listed below may be found on the Regents exam.

CONGRUENCE G-CO

Experiment with Transformations in the Plane

1. Know precise definitions of angle, circle, perpendicular line, parallel line, and line segment, based on the undefined notions of point, line, distance along a line, and distance around a circular arc.

2. Represent transformations in the plane using, for example, transparencies and geometry software; describe transformations as functions that take points in the plane as inputs and give other points as outputs. Compare transformations that preserve distance and angle to those that do not (e.g., translation versus horizontal stretch).

3. Given a rectangle, parallelogram, trapezoid, or regular polygon, describe the rotations and reflections that carry it onto itself.

4. Develop definitions of rotations, reflections, and translations in terms of angles, circles, perpendicular lines, parallel lines, and line segments.

5. Given a geometric figure and a rotation, reflection, or translation, draw the transformed figure using, for example, graph paper, tracing paper, or geometry software. Specify a sequence of transformations that will carry a given figure onto another.

Understand Congruence in Terms of Rigid Motions

6. Use geometric descriptions of rigid motions to transform figures and to predict the effect of a given rigid motion on a given figure; given two figures, use the definition of congruence in terms of rigid motions to decide if they are congruent.

7. Use the definition of congruence in terms of rigid motions to show that two triangles are congruent if and only if corresponding pairs of sides and corresponding pairs of angles are congruent.

8. Explain how the criteria for triangle congruence (ASA, SAS, and SSS) follow from the definition of congruence in terms of rigid motions.

Prove Geometric Theorems

9. Prove theorems about lines and angles. *Theorems include: vertical angles are congruent; when a transversal crosses parallel lines, alternate interior angles are congruent and corresponding angles are congruent; points on a perpendicular bisector of a line segment are exactly those equidistant from the segment's endpoints.*

10. Prove theorems about triangles. *Theorems include: measures of interior angles of a triangle sum to 180°; base angles of isosceles triangles are congruent; the segment joining midpoints of two sides of a triangle is parallel to the third side and half the length; the medians of a triangle meet at a point.*

11. Prove theorems about parallelograms. *Theorems include: opposite sides are congruent; opposite angles are congruent; the diagonals of a parallelogram bisect each other; and conversely, rectangles are parallelograms with congruent diagonals.*

Make Geometric Constructions

12. Make formal geometric constructions with a variety of tools and methods (compass and straightedge, string, reflective devices, paper folding, dynamic geometric software, etc.). *Copying a segment; copying an angle; bisecting a segment; bisecting an angle; constructing perpendicular lines, including the perpendicular bisector of a line segment; and constructing a line parallel to a given line through a point not on the line.*

13. Construct an equilateral triangle, a square, and a regular hexagon inscribed in a circle.

SIMILARITY, RIGHT TRIANGLES, AND TRIGONOMETRY G-SRT

Understand Similarity in Terms of Similarity Transformations

1. Verify experimentally the properties of dilations given by a center and a scale factor:

   a. A dilation takes a line not passing through the center of the dilation to a parallel line and leaves a line passing through the center unchanged.

   b. The dilation of a line segment is longer or shorter in the ratio given by the scale factor.

2. Given two figures, use the definition of similarity in terms of similarity transformations to decide if they are similar; explain using similarity transformations the meaning of similarity for triangles as the equality of all corresponding pairs of angles and the proportionality of all corresponding pairs of sides.

3. Use the properties of similarity transformations to establish the AA criterion for two triangles to be similar.

Prove Theorems Involving Similarity

4. Prove theorems about triangles. *Theorems include: a line parallel to one side of a triangle divides the other two proportionally and conversely; the Pythagorean theorem proved using triangle similarity.*

5. Use congruence and similarity criteria for triangles to solve problems and to prove relationships in geometric figures.

Define Trigonometric Ratios and Solve Problems Involving Right Triangles

6. Understand that by similarity, side ratios in right triangles are properties of the angles in the triangle, leading to definitions of trigonometric ratios for acute angles.

7. Explain and use the relationship between the sine and cosine of complementary angles.

8. Use trigonometric ratios and the Pythagorean theorem to solve right triangles in applied problems.

CIRCLES G-C

Understand and Apply Theorems About Circles

1. Prove that all circles are similar.

2. Identify and describe relationships among inscribed angles, radii, and chords. *Include the relationship between central, inscribed, and circumscribed angles; inscribed angles on a diameter as right angles; the radius of a circle being perpendicular to the tangent where the radius intersects the circle.*

3. Construct the inscribed and circumscribed circles of a triangle, and prove properties of angles for a quadrilateral inscribed in a circle.

Find Arc Lengths and Areas of Sectors of Circles

4. Derive using similarity the fact that the length of the arc intercepted by an angle is proportional to the radius, and define the radian measure of the angle as the constant of proportionality; derive the formula for the area of a sector.

EXPRESSING GEOMETRIC PROPERTIES WITH EQUATIONS G-GPE

Translate Between the Geometric Description and the Equation for a Conic Section

1. Derive the equation of a circle of given center and radius using the Pythagorean theorem; complete the square to find the center and radius of a circle given by an equation.

2. Derive the equation of a parabola given a focus and directrix.

Use Coordinates to Prove Simple Geometric Theorems Algebraically

3. Use coordinates to prove simple geometric theorems algebraically. *For example, prove or disprove that a figure defined by four given points in the coordinate plane is a rectangle; prove or disprove that the point $(1, \sqrt{3})$ lies on the circle centered at the origin and containing the point $(0, 2)$.*

4. Prove the slope criteria for parallel and perpendicular lines and use them to solve geometric problems (e.g., find the equation of a line parallel or perpendicular to a given line that passes through a given point).

5. Find the point on a directed line segment between two given points that partitions the segment in a given ratio.

6. Use coordinates to compute perimeters of polygons and areas of triangles and rectangles (e.g., using the distance formula).

GEOMETRIC MEASUREMENT AND DIMENSION G-GMD

Explain Volume Formulas and Use Them to Solve Problems

1. Give an informal argument for the formulas for the circumference of a circle, area of a circle, volume of a cylinder, pyramid, and cone. *Use dissection arguments, Cavalieri's principle, and informal limit arguments.*

2. Use volume formulas for cylinders, pyramids, cones, and spheres to solve problems.

Visualize Relationships Between Two-dimensional and Three-dimensional Objects

3. Identify the shapes of 2-dimensional cross-sections of 3-dimensional objects, and identify 3-dimensional objects generated by rotations of 2-dimensional objects.

MODELING WITH GEOMETRY G-MG

Apply Geometric Concepts in Modeling Situations

1. Use geometric shapes, their measures, and their properties to describe objects (e.g., modeling a tree trunk or a human torso as a cylinder).

2. Apply concepts of density based on area and volume in modeling situations (e.g., persons per square mile, BTUs per cubic foot).

3. Apply geometric methods to solve design problems (e.g., designing an object or structure to satisfy physical constraints or minimize cost; working with typographic grid systems based on ratios).

NOTES

NOTES

NOTES

NOTES

NOTES

NOTES